Nadia Nedjah, Leandro dos Santos Coelho, Luiza de Macedo Mourelle (Eds.)

Mobile Robots: The Evolutionary Approach

Studies in Computational Intelligence, Volume 50

Editor-in-chief
Prof. Janusz Kacprzyk
Systems Research Institute
Polish Academy of Sciences
ul. Newelska 6
01-447 Warsaw
Poland
E-mail: kacprzyk@ibspan.waw.pl

Further volumes of this series
can be found on our homepage:
springer.com

Nadia Nedjah
Leandro dos Santos Coelho
Luiza de Macedo Mourelle
(Eds.)

Mobile Robots: The Evolutionary Approach

With 108 Figures and 19 Tables

Springer

Nadia Nedjah
Universidade do Estado do Rio de Janeiro
Faculdade de Engenharia
Rua São Francisco Xavier
20550–900, Maracanã
524, sala 5022-D
Brazil
E-mail: nadia@eng.uerj.br

Luiza de Macedo Mourelle
Universidade do Estado do Rio de Janeiro
Faculdade de Engenharia
Rua São Francisco Xavier
20550–900, Maracanã
524, sala 5022-D
Brazil
E-mail: ldmm@eng.uerj.br

Leandro dos Santos Coelho
Pontifical Catholic University of Paraná
Rua Imaculada Conceição
1155–Prado Velho
80215–901, Curitiba, Paraná
Brazil
E-mail: leandro.coelho@pucpr.br

ISBN 978-3-642-08063-0 e-ISBN 978-3-540-49720-2

ISSN print edition: 1860-949X
ISSN electronic edition: 1860-9503

Springer is a part of Springer Science+Business Media
springer.com

Softcover reprint of the hardcover 1st edition 2007

Cover design: deblik, Berlin

Preface

Mobile robotic is a recent field that has roots in many engineering and science disciplines such as mechanical, electrical, mechatronics, cognitive and social sciences just to name few. A mobile robot needs efficient mechanisms of locomotion, kinematics, sensors data, localization, planning and navigation that enable it to travel throughout its environment. Scientists have been fascinated by conception of mobile robots for many years. Machines have been designed with wheels and tracks or other locomotion devices and/or limbs to propel the unit. When the environment is well ordered these machines can function well. Mobile robots have demonstrated strongly their ability to carry out useful work.

Intelligent robots have become the focus of intensive research in the last decade. The field of intelligent mobile robotics involves simulations and real-world implementations of robots which adapt themselves to their partially unknown, unpredictable and sometimes dynamic environments.

The design and control of autonomous intelligent mobile robotic systems operating in unstructured changing environments includes many objective difficulties. There are several studies about the ways in which, robots exhibiting some degree of autonomy, adapt themselves to fit in their environments. The application and use of bio-inspired techniques such as reinforcement learning, artificial neural networks, evolutionary computation, swarm intelligence and fuzzy systems in the design and improvement of robot designs is an emergent research topic. Researchers have obtained robots that display an amazing slew of behaviours and perform a multitude of tasks. These include, but are not limited to, perception of environment, localisation, walking, planning and navigation in rough terrain, pushing boxes, negotiating an obstacle course, playing ball, plant inspection, transportation systems, control systems for rescue operations, foraging strategies and design of automatic guided vehicles in manufacturing plants.

In this context, mobile robots designed using evolutionary computation approaches, usually known as *mobile evolutionary robotics*, have experienced significant development in the last decade. The fundamental goal of mobile evolutionary robotics is to apply evolutionary computation methods such as genetic algorithms, genetic programming, evolution strategies, evolutionary programming and differential evolution to automate the production of complex behavioural robotic controllers.

This volume offers a wide spectrum of sample works developed in leading research throughout the world about evolutionary mobile robotics and demonstrates the success of the technique in evolving efficient and capable mobile robots. The book should be useful both for beginners and experienced researchers in the field of mobile robotics. In the following, we go through the main content of the chapter included in this volume, which is organised in two main parts: Evolutionary Mobile Robots and Learning Mobile Robots

Part I. Evolutionary Mobile Robots

In Chapter 1, which is entitled *Differential Evolution Approach Using Chaotic Sequences Applied to Planning of Mobile Robot in a Static Environment with Obstacles*, the authors introduce a new hybrid approach of differential evolution combined with chaos (DEC) to the optimization for path planning of mobile robots. The new chaotic operators are based on logistic map with exponential decreasing, and cosinoidal decreasing. They describe and evaluate two case studies of static environment with obstacles. Using simulation results, the authors show the performance of the DEC in different environments in the planned trajectories. They also compared the results of DEC with classical differential evolution approaches. From the simulation results, The authors observed that the convergence speed of DEC is better than classical differential evolution. They claim that the simplicity and robustness of DEC, in particular, suggest their great utility for the problem's path planning in mobile robotics, as well as for other optimization-related problems in engineering.

In Chapter 2, which is entitled *Evolving Modular Robots for Rough Terrain Exploration*, the authors propose an original method for the evolutionary design of robotic systems for locomotion on rough terrain. They encompass the design of wheeled, legged or hybrid robots for their wide range of capabilities for support and propulsion. Their goal is to optimize the mechanical and the control system to achieve a locomotion task in a complex environment (irregular, sliding or even with uncertainties). They guarantee that the modular approach brings the possibility to match the diversity of tasks with the combination of assembly modes and that this global approach embeds an evolutionary algorithm with a dynamic simulation of the mobile robot operating in its environment. The authors claim that the hybrid encoding of the genotype allows evolving the robot morphology and its behaviour simultaneously.

They also propose specialized genetic operators to manipulate this specific encoding and to maintain their efficiency through evolution.

In Chapter 3, which is entitled *Evolutionary Navigation of Autonomous Robots Under Varying Terrain Conditions*, the authors present a fuzzy-genetic approach that provides both path and trajectory planning, and has the advantage of considering diverse terrain conditions when determining the optimal path. They modeled the terrain conditions using fuzzy linguistic variables to allow for the imprecision and uncertainty of the terrain data. The authors claim that although a number of methods have been proposed using GAs, few are appropriate for a dynamic environment or provide response in real-time. They guarantee that the proposed method is robust, allowing the robot to adapt to dynamic conditions in the environment.

In Chapter 4, which is entitled *Aggregate Selection in Evolutionary Robotics*, the authors investigate how aggregate fitness functions have been and continue to be used in evolutionary robotics, what levels of success they have generated relative to other fitness measurement methods, and how problems with them might be overcome.

In Chapter 5, which is entitled *Evolving Fuzzy Classifier for Novelty Detection and Landmark Recognition by Mobile Robots*, the authors present an approach to real-time landmark recognition and simultaneous classifier design for mobile robotics. The approach is based on the recently developed evolving fuzzy systems (EFS) method, which is based on subtractive clustering method and its on-line evolving extension called eClustering. The authors propose a novel algorithm that is recursive, non-iterative, incremental and thus computationally light and suitable for real-time applications. They report experiments carried out in an indoor environment (an office located at InfoLab21, Lancaster University, Lancaster, UK) using a Pioneer3 DX mobile robotic platform equipped with sonar and they introduce and analyse motion sensors as a case study. The authors also suggest several ways to use the engineered algorithm.

Part II. Learning Mobile Robots

In Chapter 6, which is entitled *Reinforcement Learning for Autonomous Robotic Fish*, the authors discuss applications of reinforcement learning in an autonomous robotic fish, called Aifi. They develop a three-layer architecture to control it. The bottom layer consists of several primary swim patterns. The authors use a sample-based policy gradient learning algorithm in this bottom layer to evolve swim patterns. The middle layer consists of a group of behaviours which are designed for specific tasks. They apply a state-based reinforcement learning algorithm, Q-learning in particular, in the top layer to find an optimal planning policy for a specific task. They claim that both simulated and real experiments show good feasibility and performance of the proposed learning algorithms.

In Chapter 7, which is entitled *Module-based Autonomous Learning for Mobile Robots*, the author implement a solution that uses qualitative and quantitative knowledge to make robot tasks able to be treated by Reinforcement Learning (RL) algorithms. The steps of this procedure include a decomposition of the overall task into smaller ones, using abstractions and macro-operators, thus achieving a discrete action space; the application of a state model representation to achieve both time and state space discretisation; the use of quantitative knowledge to design controllers that are able to solve the subtasks; learning the coordination of these behaviours using RL, more specifically Q-learning. The authors use and evaluate the proposed method on a set of robot tasks using a Khepera robot simulator. They test two approaches for state space discretisation were tested, one based on features, which are observation functions of the environment and the other on states. They compare the learned policies over these two models to a predefined hand-crafted policy. The authors claim that the resulting compact representation allows the learning method to be applied over the state-based model, although the learned policy over the feature-based representation has a better performance.

In Chapter 8, which is entitled *A Hybrid Adaptive Architecture for Mobile Robots Based on Reactive Behaviours*, the author first describe the high-level schemas commonly adopted for intelligent agent architectures, focusing on their constituent structural elements. Then they present the main organisation of the proposed architecture, where the behaviours used and the coordination layer are respectively explained. They also describe and analyse the experiments conducted with AAREACT and the obtained results.

In Chapter 9, which is entitled *Collaborative Robots for Infrastructure Security Applications*, the author addresses the scenario of a team of mobile robots working cooperatively by first presenting distributed sensing algorithms for robot localisation and 3D map building. They also present a multi-robot motion planning algorithm according to a patrolling and threat response scenario. The authors use neural network based methods for planning a complete coverage patrolling path.

In Chapter 10, which is entitled *Imitation Learning: An Application in a Micro Robot Soccer Game*, the authors present a robot soccer system that learns by imitation and by experience. They use both learning by imitation then learning by experience as a strategy to make robots grasp the way they should play soccer. The authors claim that repeating this process allows that robots can continuously improve their performance.

We are very much grateful to the authors of this volume and to the reviewers for their tremendous service by critically reviewing the chapters. The editors would also like to thank Prof. Janusz Kacprzyk, the editor-in-chief of the Studies in Computational Intelligence Book Series and Dr. Thomas Ditzinger from Springer-Verlag, Germany for their editorial assistance and excellent collaboration to produce this scientific work. We hope that the reader will share our excitement on this volume and will find it useful.

March 2006

Nadia Nedjah, State University of Rio de Janeiro, Brazil
Leandro S. Coelho, Pontifical Catholic University of Parana, Brazil
Luiza M. Mourelle, State University of Rio de Janeiro, Brazil

Contents

List of Figures

List of Tables

Part I

Evolutionary Mobile Robots

1

Differential Evolution Approach Using Chaotic Sequences Applied to Planning of Mobile Robot in a Static Environment with Obstacles

Leandro dos Santos Coelho[1], Nadia Nedjah[2], and Luiza de Macedo Mourelle[3]

[1] Production and Systems Engineering Graduate Program
Pontifical Catholic University of Parana, LAS/PPGEPS/PUCPR
Imaculada Conceicao, 1155, Zip code 80215-901, Curitiba, Parana, Brazil
leandro.coelho@pucpr.br

[2] Department of Electronics Engineering and Telecommunications,
Engineering Faculty,
State University of Rio de Janeiro,
Rua São Francisco Xavier, 524, Sala 5022-D,
Maracanã, Rio de Janeiro, Brazil
nadia@eng.uerj.br, http://www.eng.uerj.br/~nadia

[3] Department of System Engineering and Computation,
Engineering Faculty,
State University of Rio de Janeiro,
Rua São Francisco Xavier, 524, Sala 5022-D,
Maracanã, Rio de Janeiro, Brazil
ldmm@eng.uerj.br, http://www.eng.uerj.br/~ldmm

Evolutionary algorithms have demonstrated excellent results for many engineering optimization problems. In other way, recently, the chaos theory concepts and chaotic times series have gained much attention during this decade for the design of stochastic search algorithms. Differential evolution is a new evolutionary algorithm mainly having three advantages: finds the global minimum regardless of the initial parameter values, fast convergence and uses few control parameters. In this work, a new hybrid approach of Differential Evolution combined with Chaos (DEC) is presented for the optimization for path planning of mobile robots. The new chaotic operators are based on logistic map with exponential and cosinoidal decreasing. Two case studies of static environment with obstacles are described and evaluated. Simulation results show the performance of the DEC in different environments for the planned trajectories. Results of DEC were also compared with classical differential evolution approaches. From the simulation results, it was observed

L. dos Santos Coelho et al.: *Differential Evolution Approach Using Chaotic Sequences Applied to Planning of Mobile Robot in a Static Environment with Obstacles*, Studies in Computational Intelligence (SCI) **50**, 3–22 (2007)
www.springerlink.com

that the convergence speed of DEC is better than classical differential evolution. The simplicity and robustness of DEC, in particular, suggest their great utility for the problem's path planning in mobile robotics, as well as for other optimization-related problems in engineering.

1.1 Introduction

Autonomous path planning system design for intelligent robots is a long cherished goal to robotic or control engineers. One of the issues of the research on a mobile robot is to move the robot without a collision from a starting position to a goal position in the navigation space. The main methods of path planning are the cell decomposition [21, 24], potential fields [17, 12, 1], roadmaps [16, 6, 18] using Voronoi diagrams [41, 27, 28], probabilistic roadmaps [5, 13] and visibility graph [19, 14].

Recently, Evolutionary Algorithms (EAs) emerged as a powerful method to achieve goals in path planning for mobile robotics. EAs are general-purpose methods for optimization belonging to a class of meta-heuristics inspired by the evolution of living beings and genetics [9, 3]. EAs usually do not require deep mathematical knowledge of the problem and do not guarantee the optimal solution in a finite time. However, they are useful for large-scale optimization problems, dealing efficiently with huge and irregular search spaces. EAs use a population of structures (individuals), where each one is a candidate solution for the optimization problem. Since they are population-based methods, they do a parallel search of the space of possible solutions, and are less susceptive to local minima. Therefore, EAs are suited for solving a broad range of complex problems, characterized by discontinuity, non-linearity and multivariability. The usefulness of a given solution is obtained from the environment by means of a fitness function. The population of solutions evolves throughout generations, based on probabilistic transitions using cooperation and auto-adaptation of individuals. There are many variants of EAs, but the main differences rely on: how individuals are represented, the genetic operators that modify individuals (especially mutation and crossover) and the selection procedure. Most current approaches of EAs descend from principles of main methodologies: genetic algorithm (GA), evolutionary programming, evolution strategy, and differential evolution (DE).

In the literature, several authors have proposed the path planning, cooperation among robots, and design of control systems in mobile robotics by EAs. Xiao et al. developed in [46] an adaptive evolutionary planner/navigator for mobile robotics. Hocao textasciibreveglu and Sanderson presented in [11] a new approach to multi-dimensional path planning based on optimization by GA. Sipper presented in [38] a genetic algorithm to shape an environment for a simulated Khepera robot. Watanabe and Izumi provided in [44] a discussion on control and motion planning approaches to robotic manipulators or mobile robots with soft computing techniques, such as fuzzy systems, neural networks and GAs. Santos et al. showed in [33] different aspects of the use of evolution for successful generation of real robots using neural networks and GAs. Watson et al. dealt in [45] with a powerful method for evolutionary robotics using a population of physical robots that autonomously reproduce with one another while situated in their task environment. Nojima et al. showed in Nojima-2003 a human-friendly trajectory generation using an interactive GA for a partner robot. Kamei and Ishikawa presented in [15] an approach based on reinforcement learning and genetic algorithms for path planning for autonomous mobile robots.

The contribution of this paper is to present a new hybrid approach of differential evolution combined with chaos theory (DEC) for the optimization of path planning of mobile robots. The new chaotic operators are based on logistic map: logistic map with exponential decreasing and logistic cosinoidal decreasing. Two case studies of static environment with obstacles are described and evaluated. Simulation results show the performance of the DEC in different environments in the planned trajectories. Simulation results of DEC are also compared with classical differential evolution (DE) and GA approaches.

This Chapter is organized as follows: In Section 1.2, the fundamentals of differential evolution are presented. The details of the new approach of DE with chaos theory are discussed in Section 1.3. Two case studies of path planning for mobile robots are proposed in Section 1.4. The simulation results and conclusions are presented in Section 1.5 and Section 1.6, respectively.

1.2 Differential Evolution

Differential Evolution (DE) is a population-based and stochastic function minimizer (or maximizer), whose simple, yet powerful, and straightforward features make it very attractive for numerical optimization. DE uses a rather greedy and less stochastic approach to problem solving compared to EAs. DE combines simple arithmetic operators with the classical operators of crossover, mutation and selection to evolve from a randomly generated starting population to a final solution.

In [40], Storn and Price first introduced the DE algorithm a few years ago. The DE was successfully applied to the optimization of some well-known nonlinear, non-differentiable and non-convex functions in [39]. DE is an approach for the treatment of real-valued optimization problems. In this case [20], Krink et al. mentioned also that DE is a very powerful heuristic for non-noisy optimization problems, but that noise is indeed a serious problem for conventional DE, when the fitness of candidate solutions approaches the fitness variance caused by the noise.

DE is similar to a (μ, λ) evolution strategies [3], but in DE the mutation is not done via some separately defined probability density function. DE is also characterized by the use of a population-derived noise to adapt the mutation rate of the evolution process, implementation simplicity and speed of operation.

There are many variants of DE approaches that have been reported [40]. The different variants are classified using the following notation: DE/α/β/δ, where α indicates the method for selecting the parent chromosome that will form the base of the mutated vector, β indicates the number of difference vectors used to perturb the base chromosome, and δ indicates the crossover mechanism used to create the child population. The bin acronym indicates that crossover is controlled by a series of independent binomial experiments.

The fundamental idea behind DE is a scheme by which it generates the trial parameter vectors. At each time step, DE mutates vectors by adding weighted, random vector differentials to them. If the cost of the trial vector is better than that of the target, the target vector is replaced by the trial vector in the next generation. The variant implemented in this chapter is the DE/*rand*/1/*bin* and given by the following steps:

1. Initialize a population of individuals (solution vectors) with random values generated according to a uniform probability distribution in the n dimensional problem space.
2. For each individual, evaluate its fitness value.
3. Mutate individuals according to (1.1):

$$z_{i+1}(t) = x_{i,r_1}(t) + f_m \lfloor x_{i,r_2}(t) - x_{i,r_3}(t) \rfloor \tag{1.1}$$

4. Following the mutation operation, crossover is applied in the population. For each mutant vector, $z_i(t+1)$, an index $rnbr(i) \in \{1, 2, \ldots, n\}$ is randomly chosen using uniform distribution, and a *trial vector*, $u_i(t+1) = [u_{i_1}(t+1), u_{i_2}(t+1), \ldots, u_{i_n}(t+1)]^T$, is generated as described in (1.2). To decide whether or not the vector $u_i(t+1)$ should be a member of the population comprising the next generation, it is compared to the corresponding vector $x_i(t)$. Thus, if F_c denotes the objective function under minimization, then we have (1.3)

$$u_{i_j}(t+1) = \begin{cases} z_{i_j}(t+1) \text{ if } randb(j) \leq CR)\text{or } (j = rndr(i)) \\ \\ x_{i_j}(t+1) \text{ if } randb(j) > CR)\text{or } (j \neq rndr(i)) \end{cases} \tag{1.2}$$

$$x_i(t+1) = \begin{cases} u_{i_j}(t+1) \text{ if } F_c(t+1) < F_c(x_i(t)) \\ \\ x_i(t) \qquad \text{otherwise} \end{cases} \tag{1.3}$$

5. Loop to step (2) until a stopping criterion is met, usually a maximum number of iterations (generations), t_{max}.

In the above equations, $i = 1, 2, \ldots, N$ is the individual's index of population, $j = 1, 2, \ldots, n$ is the position in n dimensional individual; t is the time (generation); $x_i(t) = [x_{i_1}(t), x_{i_2}(t), \ldots, x_{i_n}(t)]^T$ stands for the position of the i-th individual of population of N real-valued n-dimensional vectors; $z_i(t) = [z_{i_1}(t), z_{i_2}(t), \ldots, z_{i_n}(t)]^T$ stands for the position of the i-th individual of a *mutant vector*; r_1, r_2 and r_3 are mutually different integers and also different from the running index, i, randomly selected with uniform distribution from the set $\{1, 2, \ldots, i-1, i+1, \ldots, N\}$; $f_m > 0$ is a real parameter, called *mutation factor*, which controls the amplification of the difference between two individuals so as to avoid search stagnation and it is usually taken from the range [0.1, 1]; $randb(j)$ is the j-th evaluation of a uniform random number generation with [0, 1]; CR is a *crossover rate* in the range [0, 1]; and Fc is the evaluation of cost function. Usually, the performance of a DE algorithm depends on three variables: the population size N, the mutation factor f_m, and the crossover rate CR.

1.3 New Approach of Differential Evolution Combined with Chaos Theory

The chaos theory studies the unexpected phenomenon apparently, in the search of hidden standards and simple laws that conduct the complex behaviors. However, this study became effectively reasonable from the decade of 1960, when the computers had started to possess reasonable graphical and processing capacities, giving to the physicists and mathematicians the power to discover answers for basic questions of the science in general way, that were obscure before.

The behavior of chaotic systems presents great sensitivity in relation to the initial conditions that are applied. Chaos, apparently disordered behavior that is nonetheless deterministic, is a universal phenomenon that occurs in many systems in all areas of science. The randomness of chaotic sequences is a result of the sensitivity of chaotic systems to the initial conditions. However, because the systems are deterministic, chaos implies some order. A system can make the transformation from a regular periodic system to a chaotic system simply by altering one of the controlling parameters [35].

The nonlinear systems had appeared from the chaos theory that supplies to an explanation, many times adequate, to many behaviors current in biological systems [2], such as: natural phenomenon (populations, turbulence, fluid movement, and cloud formation), complexities in electric circuits [23], telecommunications [34], control systems, dynamic behavior of the cardiac beatings [32], among others.

The chaos theory has received growing attention, with few papers study application of chaos in optimization methods [22, 26, 48, 10, 47, 25].The optimization algorithms based on chaos theory are stochastic search methodologies that differ from any of the existing EAs. EAs are optimization approaches with in concepts bio-inspired of genetics and natural evolution. In other way, chaotic optimization approaches are based on ergodicity, stochastic properties and irregularity. It is not like some stochastic optimization algorithms that escape from local minima by accepting some bad solutions according to a certain probability [35]. The chaotic optimization approaches can more easily escape from local minima that can other stochastic optimization algorithms [22].

In the context of DE, the concepts of chaotic optimization methods can be useful. Generally, the parameters f_m, CR and N of DE are the key factors to affect the convergence of the algorithm. In fact, however, parameters f_m and CR cannot ensure the optimization's ergodicity entirely in phase search because they are constant factors in traditional DE. Therefore, this paper provides three new approaches that introduces chaotic mapping with ergodicity, irregularity and the stochastic property into DE to improve the global convergence. The utilization of chaotic sequences in EAs can be useful for escape more easily from local minima than the traditional EAs.

One of the simplest dynamic systems evidencing chaotic behavior is the iterator named logistic map [29], whose equation is given in (1.4)

$$y(t+1) = \mu y(t)\,(1 - y(t)) \tag{1.4}$$

where $t = 1, \ldots, t_{max}$, t_{max} is the number of samples, μ is a control parameter, $0 \le \mu \le 4$. The behavior of system of (1.4) is greatly changed with the variation of μ. Its value determines whether y stabilizes at a constant size, oscillates between a limited sequence of sizes, or whether y behaves chaotically in an unpredictable pattern. A very small difference in the initial value of y causes large difference in its long-time behavior [47]. Equation (1.4) is deterministic, it exhibits chaotic dynamics when $\mu = 4$ and $y(1) \ni \{0, 0.25, 0.50, 0.75, 1\}$. In this case, $y(t)$ is distributed in the range (0,1) under the conditions that the initial $y(1) \in (0,1)$ and that $y(1) \ni \{0, 0.25, 0.50, 0.75, 1\}$.

The new approaches of DE combined with chaos (DEC) based on logistic maps and new variants based on exponential and cosinoidal functions are described as follows:

Approach 1 – DEC(1). The parameter f_m of (1.1) is modified using (1.4) through (1.5), (1.6) and (1.7):

$$z_{i+1}(t) = x_{i,r_1}(t) + f_m(t)\lfloor x_{i,r_2}(t) - x_{i,r_3}(t)\rfloor \tag{1.5}$$

$$f_m(t+1) = \mu f_m(t)\,(1 - f_m(t)) \tag{1.6}$$

$$f_m(t) \in (0,1) \tag{1.7}$$

Approach 2 – DEC(2). The parameter f_m of (1.1) is modified by (1.4) using (1.8):

$$z_{i+1}(t) = x_{i,r_1}(t) + f_m(t)\lfloor x_{i,r_2}(t) - x_{i,r_3}(t)\rfloor \tag{1.8}$$

$$f_m(t) = \left| e^{-\frac{t}{3t_{max}}} \right| (\mu f_m(t)\,[1 - f_m(t)]) \tag{1.9}$$

$$f_m(t) \in (0, 1) \tag{1.10}$$

where t is the current generation and t_{max} is the maximum number of iterations (generations) of optimization procedure.

Approach 3 – DEC(3). The parameter f_m of (1.1) is modified by (1.4) as in (1.11).

$$z_{i+1}(t) = x_{i,r_1}(t) + f_m(t)\lfloor x_{i,r_2}(t) - x_{i,r_3}(t)\rfloor \tag{1.11}$$

$$f_m(t+1) = 0.8\,|\sin 6t|\,(\mu f_m(t)\,[1 - f_m(t)]) + 0.4 \tag{1.12}$$

$$f_m(t) \in (0, 1) \tag{1.13}$$

Approach 4 – DEC(4). The parameter CR of (1.2) is modified by (1.4).

$$u_{i_j}(t+1) = \begin{cases} z_{i_j}(t+1) \text{ if } randb(j) \le CR(t))\text{or } (j = rndr(i)) \\ x_{i_j}(t+1) \text{ if } randb(j) > CR(t))\text{or } (j \ne rndr(i)) \end{cases} \tag{1.14}$$

$$CR(t+1) = \mu CR(t)\,(1 - CR(t)) \tag{1.15}$$

$$CR(t) \in (0, 1] \tag{1.16}$$

Approach 5 – DEC(5). This approach uses simultaneously the DEC(1) and the DEC(4) in optimization procedure.

Approach 6 – DEC(6). This approach uses simultaneously the DEC(2) and the DEC(4) in optimization procedure.

Approach 7 – DEC(7). This approach uses simultaneously the DEC(3) and the DEC(4) in optimization procedure.

The following values for the parameters $CR(1)$ and $f_m(1)$ to 0.48 and $\mu = 4$ have been used for simulation with DEC approaches. In this case, the track of chaotic variable can travel ergodicaly over the whole search space. In Figure 1.1, Figure 1.2 and Figure 1.3 and, Figure 1.4, Figure 1.5 and Figure 1.6, the distribution of points in DE(2), DE(3) and DE(4) and, DEC(1), DEC(2) and DEC(3) are presented, respectively. Figure 1.4, Figure 1.5 and Figure 1.6 show that a lot of points of logistic maps and variants distribute near the edges useful for DEC approaches escape of local optimum.

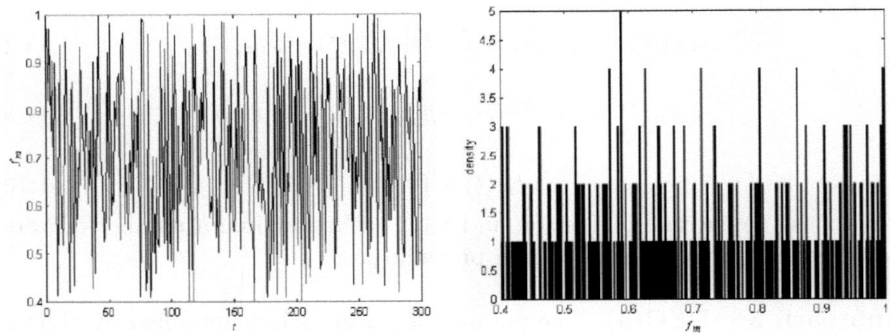

(a) function $f_m(t)$ (b) Probabilistic density function

Fig. 1.1. Illustration for the case of DE(2)

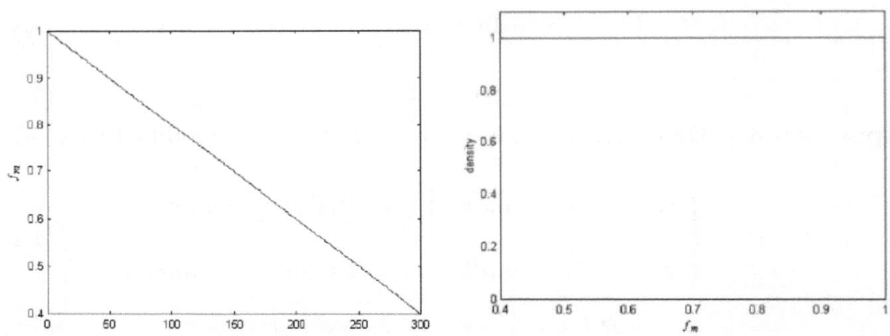

(a) function $f_m(t)$ (b) Probabilistic density function

Fig. 1.2. Illustration for the case of DE(3) using linear decreasing

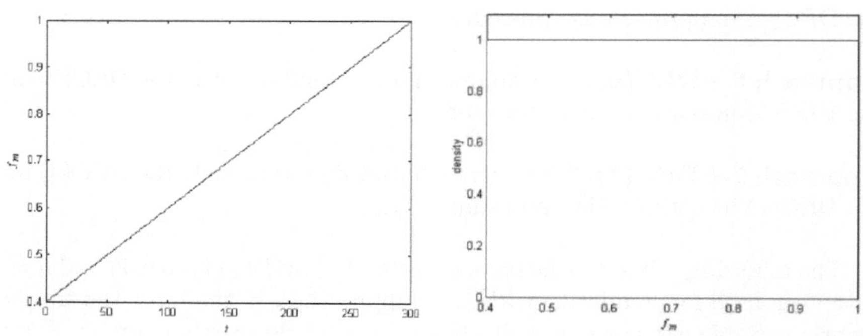

(a) function $f_m(t)$ (b) Probabilistic density function

Fig. 1.3. Illustration for the case of DE(4) with linear increasing

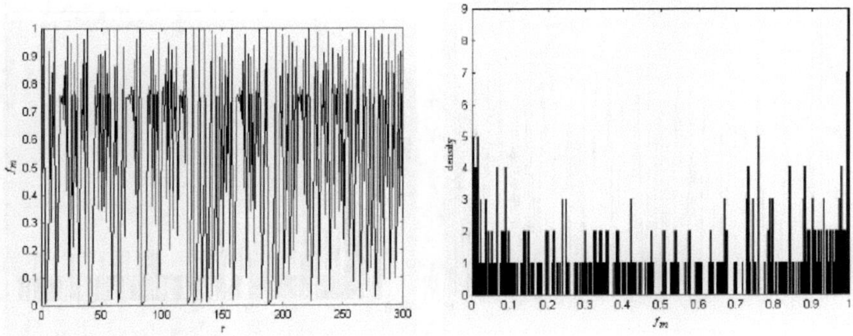

(a) function $f_m(t)$ (b) Probabilistic density function

Fig. 1.4. Illustration for the case of DEC(1)

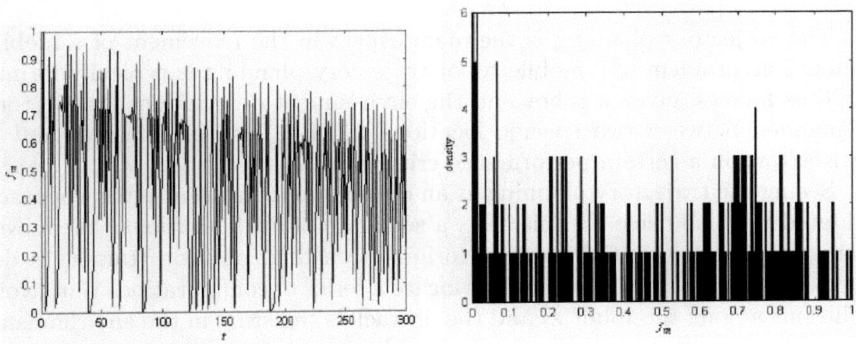

(a) function $f_m(t)$ (b) Probabilistic density function

Fig. 1.5. Illustration for the case of DEC(2)

1.4 Planning of Mobile Robots

The literature is rich in approaches to solve mobile robots trajectory planning in presence of static and/or dynamic obstacles [43, 4, 30, 36, 37]. One of the most popular planning methods is the artificial potential field [42]. However, this method gives only one trajectory solution that may not be the smaller trajectory in a static environment. The main difficulties in determining the optimum trajectory are due to the fact that many analytical methods are complex to be used in real time, and the searching enumerative methods are excessively affected by the size of the searching space.

Recently, the interest in using EAs, especially genetic algorithms, has increased in last years. Genetic algorithms are used in mobile robots trajectory planning, generally when the search space is large [7, 46, 8].

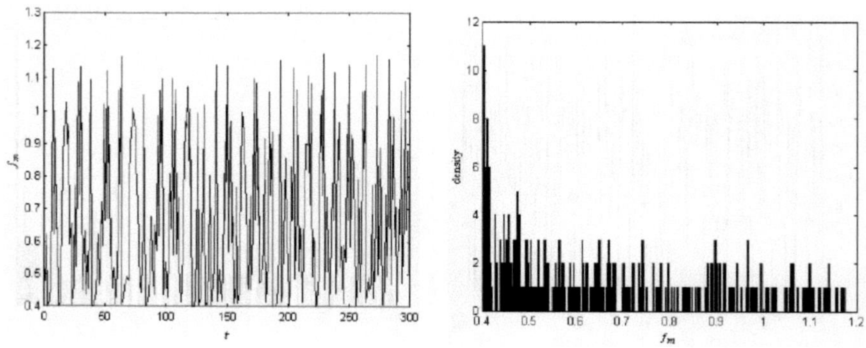

(a) function $f_m(t)$ (b) Probabilistic density function

Fig. 1.6. Illustration for the case of DEC(3)

The trajectory planning is the main aspect in the movement of a mobile robot. The problem of a mobile robot trajectory planning is typically formulated as follows: given a robot and the environment description, a trajectory is planned between two specific locations which is free of collisions and is satisfactory in a certain performance criteria [46].

Seeing the trajectory planning as an optimization problem is the approach adopted in this article. In this case, a sequence of configurations that moves the robot from an initial position (origin) to a final position (target) is designed. A trajectory optimizer must locate a series of configurations that avoid collisions among the robot(s) and the obstacle(s) existing in the environment. The optimizer must also try to minimize the trajectory length found, in order to be efficient. The search space is the group of all possible configurations.

In the present study, it's considered a two-dimensional mobile robot trajectory planning problem, in which the position of the mobile robot R is represented by Cartesian coordinates (x, y) in the xy plan. The initial and destination points of the robot are (x_0, y_0) and (x_{n_p}, y_{n_p}), where n_p is a design parameter. The initial point is always $(0, 0)$.

Only the problem's trajectory planning is empathized in this chapter, the robot control problem is not the focus here. However, details of the robots movement equations can be found in [7]. Its assumed that the obstacles are circular in the robot's moving plan. Besides, the hypothesis that the free two-dimensional space is connected and the obstacles are finite in size and does not overlap the destiny point is true.

The optimization problem formulated consists of a discrete optimization problem, where the objective function $f(x, y)$, which is the connection between the technique used for optimization and the environment, aims to minimize the total trajectory roamed by the mobile robot and is ruled as in 1.17

$$f(x,y) = \alpha d_{obj} + \lambda n_O \tag{1.17}$$

$$d_{obj} = \sum_{i=0}^{n_p} \sqrt{(x(i+1) - x(i))^2 + (y(i+1) - y(i))^2} \tag{1.18}$$

where α and λ are weighted factors, d_{obj} represents the Euclidian distance between the initial and the destiny points, n_O denotes the number of obstacles prevented by the robot movement following the planned trajectory, and n_p is the number of points where a trajectory change occurs (project parameter in this chapter). It is described in (1.17) that a λ term exists, it is a weighting or penalty term for unfeasible solutions, meaning, the trajectory that intercepts obstacles. In this case, the fitness function to be evaluated by optimization approaches of this paper aims to maximize

$$fitness = \frac{K_c}{f(x,y) + \epsilon} \tag{1.19}$$

where K_c and ϵ are scale constants.

1.5 Simulation Results

The environment used for the trajectory planning is a 100×100 meters field. The search interval of the parameters is $x_i in[0,100]$ and $y_i \in [0,100]$ meters, where $i = 1,\ldots,n_p$. About the fitness its adopted $\alpha = 1$, $\lambda = 200$, $K_c = 100$ and $\epsilon = 1 \times 10^{-6}$. The parameters of population size (N) and maximum number of generations (t_{max}) used in implementations of the DE and DEC are 30 and 100 for the two case studies, respectively. The following traditional DE were tested:

- DE(1): $CR = 0.80$ and $f_m = 0.40$;
- DE(2): $CR = 0.80$ and $f_m = 0.60 \times r + 0.40$, where r is a random number with uniform distribution in the range [0,1];
- DE(3): $CR = 0.80$ and f_m with linear decreasing, where $f_m = (f_{mf} - f_{mi})(g/t_{max}) + f_{mi}$; the constants f_{mf} and f_{mi} are 1.0 and 0.4, respectively;
- DE(4): $CR = 0.80$ and f_m with linear increasing, where $f_m = (f)mf - f_{mi})(g/t_{max}) + f_{mi}$; the constants f_{mf} and f_{mi} are 0.4 and 1.0, respectively.

In the next section, we present two simulated cases and the results analysis of 30 experiments with the DE and DEC algorithms.

1.5.1 Case study 1: Environment with 7 obstacles

In Table 1.1 are presented the positions of the centers (x_c, y_c) of the circular obstacles and their respective radius (in meters) of case 1. The results obtained with the DE and DEC are restricted to $np = 2$. In Table 1.2, the obtained

Table 1.1. Definition of 7 obstacles for the case study 1

Obstacle Number	Radius	Position (x_c, y_c)
1	15	(50, 60)
2	10	(85, 55)
3	07	(93, 80)
4	10	(15, 30)
5	10	(70, 90)
6	15	(50, 20)
7	10	(20, 90)

Table 1.2. Results for an environment with 7 obstacles for 30 runs (best results for each experiment after 300 generations)

Optimization Technique	Maximum Fitness	Mean Fitness	Minimum Fitness	Standard Deviation
DE(1)	**0.7025**	0.5976	0.2929	0.1264
DE(2)	**0.7025**	0.6117	0.5270	0.0709
DE(3)	**0.7025**	0.5735	0.5270	0.0714
DE(4)	**0.7025**	0.5699	0.2929	0.1605
DEC(1)	**0.7025**	0.6246	0.6044	**0.0410**
DEC(2)	**0.7025**	0.6172	0.5259	0.0514
DEC(3)	**0.7025**	0.6128	0.5270	0.0819
DEC(4)	**0.7025**	0.6235	0.5264	0.0728
DEC(5)	**0.7025**	**0.6736**	**0.6045**	0.0465
DEC(6)	**0.7025**	0.5919	0.5270	0.0533
DEC(7)	**0.7025**	0.6456	0.5300	0.0631

solutions and statistical analysis of results are presented, wherein the best performing algorithm(s) is highlighted.

In case study 1, all DE and DEC algorithms are able to find optimum solutions. The best (higher) fitness that the DE(1)–(4) and DEC(1)–(7) have achieved, for $np = 2$, has been obtained with the solution: $(x_1, y_1) = (87.5785; 84.4280)$ and $(x_2, y_2) = (60.8406; 49.4837)$. The classical DE(4) approach presents a minimum fitness of 0.5699. However, the DEC(6) usign chaos theory performs a minimum fitness of 0.5919. In the case study 1, DEC(5) obtains a more efficient convergency in terms of mean and minimum fitness that the others tested DE and DEC approaches.

In Figure 1.7 the best result of the experiments is presented. In case study 1, the best of DE(1)–(4) and DEC(1)–(7) obtain a distance total of path of 142.343. This distance is 99.352% of optimum path without obstacles.

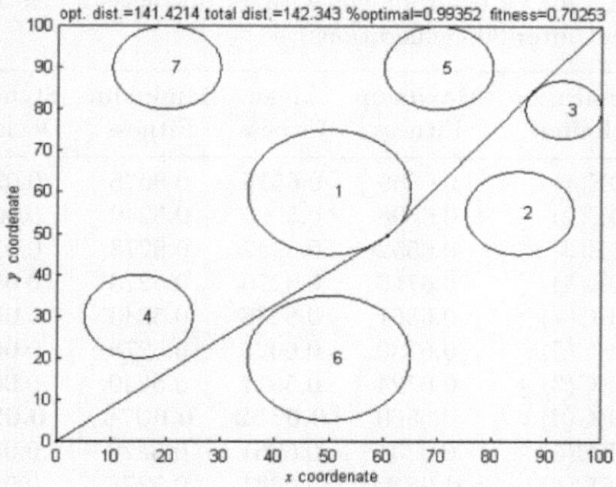

Fig. 1.7. Best result achieved for the DE(1)–(4) and DEC(1)–(7) approaches in case study 1

1.5.2 Case study 2: Environment with 14 obstacles

In Table 1.3 are presented the center positions (x_c, y_c) of the circular obstacles and their respective radius (in meters) for case 2. The results obtained for the DE and DEC are restricted to np=3. In Table 1.4, the results for the case study 2 are summarized.

Table 1.3. Obstacles for the case study 2

Obstacle Number	Radius	Position (x_c, y_c)
1	04	(95, 60)
2	15	(20, 30)
3	15	(20, 85)
4	11	(55, 50)
5	10	(50, 85)
6	08	(80, 60)
7	05	(90, 80)
8	14	(80, 20)
9	05	(65, 70)
10	05	(90, 42)
11	08	(50, 08)
12	04	(90, 95)
13	05	(60, 30)
14	04	(40, 65)

In Figure 1.8(a)–(d) and Figure 1.9(a)–(g), the best results of the experiments with DE and DEC are presented. In this case, the best result of the

Table 1.4. Results for an environment with 14 obstacles for 30 runs (best results for each experiment after 300 generations)

Optimization Technique	Maximum Fitness	Mean Fitness	Minimum Fitness	Standard Deviation
DE(1)	0.6869	0.6515	0.6075	0.0229
DE(2)	0.6598	0.5995	0.5240	0.0625
DE(3)	0.6552	0.5942	0.5278	0.0548
DE(4)	0.6716	0.6270	0.5278	0.0528
DEC(1)	0.6561	0.6236	0.5243	0.0510
DEC(2)	0.6730	0.6025	0.5278	0.0664
DEC(3)	0.6724	0.5912	0.5240	0.0687
DEC(4)	0.6869	**0.6552**	**0.6076**	**0.0227**
DEC(5)	0.6731	0.6261	0.5278	0.0589
DEC(6)	**0.6884**	0.6331	0.5278	0.0567
DEC(7)	0.6624	0.6214	0.5240	0.0516

experiments was: $(x_1, y_1) = (31.1352; 19.2243)$, $(x_2, y_2) = (63.9030; 42.8367)$ and $(x_3, y_3) = (75.9197; 70.7897)$ using DEC(6).

The mean performance would be useful as an indication of the robustness of the configuration of DE and DEC. In case study 2, the best result of DEC(6) obtains a total distance of 145.2639. This distance is 97.355% of optimum path without obstacles. However, DEC(4) presents the best mean, minimum and standard deviation of tested approaches. In Table 1.4, it is observed that the DE and DEC responded well for all the simulations attempts. From 30 repeated simulations, it is shown that the results of DEC were significant, in terms of mean fitness, for path planning in relation of DE for the case study 2. As seen in the comparative study, the robustness of the DEC is higher than the one of the DE specially when dealing with more complex environments.

1.6 Summary

A research area with special relevance to mobile robot systems is devising suitable methods to plan optimum moving trajectories. There exist many approaches within the area of EAs to solve the problem of optimization of path planning in mobile robotics. In this paper the application of a new approach of DE based on chaos theory in form of an optimization algorithm is explored.

DE algorithm is an EA approach mainly having three potential advantages; finding the true global minimum regardless of the initial parameter values, fast convergence, and using few control parameters. DEC approaches appear to be a good choice, since it does not require elaborate tuning methods for the control parameters and their performances is very reliable compared to traditional DE. In this chapter, the possibilities of exploring the DEC efficiency are successfully presented, as shown in two simulated cases study. The

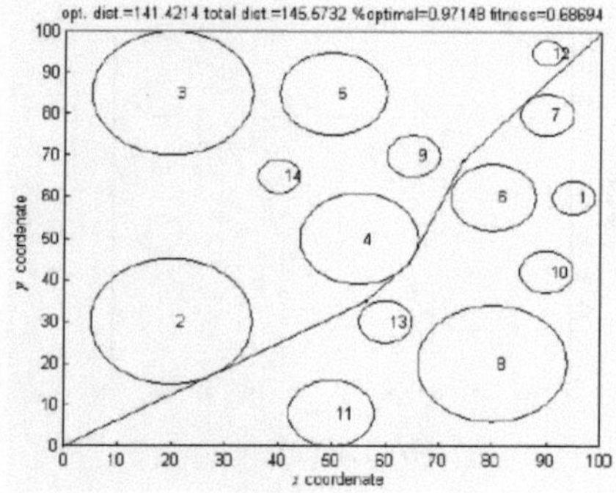

(a) DE(1)

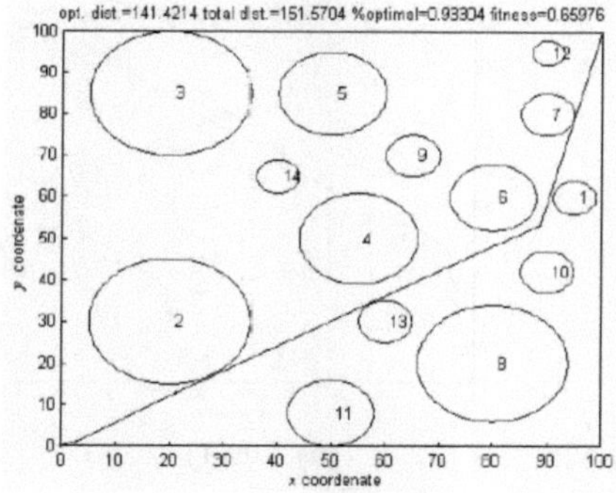

(b) DE(2)

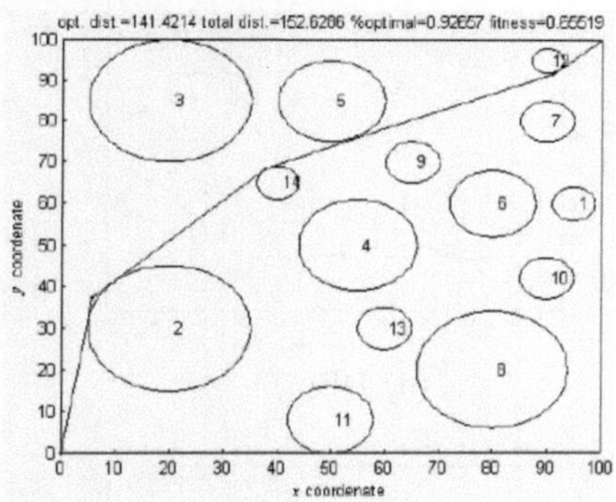

(c) DE(3)

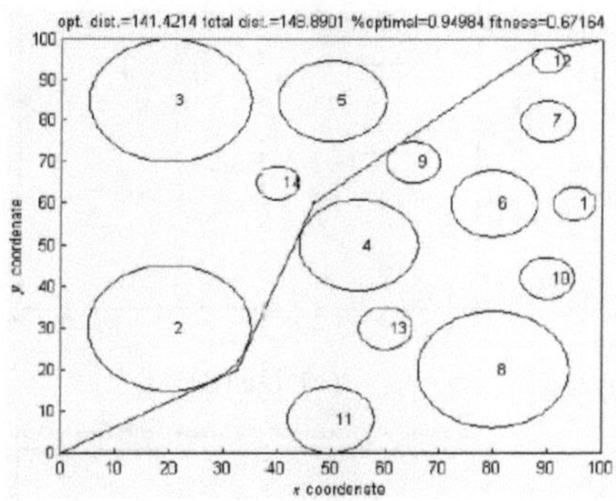

(d) DE(4)

Fig. 1.8. Best result achieved for the DE(1)–(4) in case study 2

results of these simulations are very encouraging and they indicate important contributions to the areas of setup of differential evolution algorithms and path planning in mobile robotics. DE with chaos theory is employed in this paper for enhance the global exploration of traditional DE.

Among the tested algorithms, the DEC(4), DEC(5) and DEC(6) can rightfully be regarded as a good choice due to its convergence speed and robustness in global search. However, in future works, more detailed studies and experiments related to population size and design parameters of DE and DEC are necessary. In this context, a comparative study of DEC with other EAs methodologies will be done.

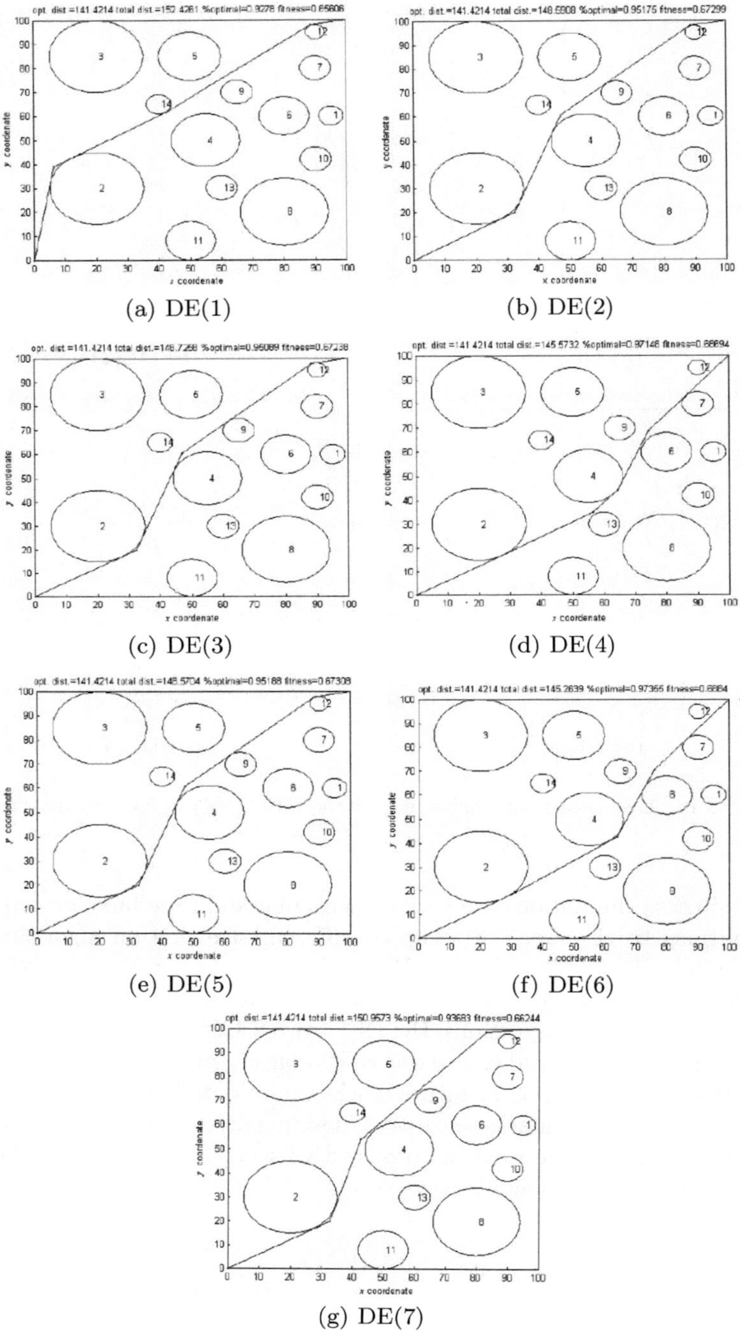

Fig. 1.9. Best result achieved for the DEC(1)–(7) in case study 2

Acknowledgments

The first author gratefully acknowledges support from Fundao Araucária of Paraná (Araucaria Foundation of Paraná state), Brazil, through project with Grant No. 6234 (005/2005).

References

1. Agirrebeitia, J., Avilés, R., Bustos, I.F., Ajuria, G., A new APF for path planning in environments with obstacles. Mechanism and Machine Theory, 2005.
2. Aguirre, L. A., Billings, S. A., Nonlinear chaotic systems: approaches and implications for science and engineering - a survey, Applied Signal Processing 2, pp. 224–248, 1995.
3. Bäck, T., Fogel, D.B., Michalewicz, Z., Handbook of evolutionary computation. Bristol, Philadelphia: Institute of Physics Publishing, New York, Oxford: Oxford University Press, 1997.
4. Bennewitz, M., Burgard, W., Thrun, S., Finding and optimizing solvable priority schemes for decoupled path planning techniques for teams of mobile robots. Robotics and Autonomous Systems 41, pp. 89–99, 2002.
5. Dale, L.K., Amato, N.M., Probabilistic roadmaps-putting it all together. IEEE International Conference on Robotics and Automation, Seoul, Korea, vol. 2, pp. 1940–1947, 2001.
6. Dudek, G., Jenkin, M., Computational principles of mobile robotics. Cambridge University Press, Cambridge, UK, 2000.
7. Fujimori, A., Nikiforuk, P.N., Gupta, M.M., Adaptive navigation of mobile robots with obstacle avoidance. IEEE Transactions on Robotics and Automation 13(4), pp. 596–602, 1997.
8. Gemeinder, M., Gerke, M., GA-based path planning for mobile robot systems employing an active search algorithm. Applied Soft Computing 3, pp. 149–158, 2003.
9. Goldberg, D.E., Genetic algorithms in search, optimization, and machine learning. Addison-Wesley, MA, 1989.
10. Hasegawa, M., Ikeguchi, T., Aihara, K., Itoh, K., A novel chaotic search for quadratic assignment problems. European Journal of Operational Research 139, pp. 543–556, 2002.
11. Hocaoglu, C., Sanderson, A.C., Planning multiple paths with evolutionary speciation. IEEE Transactions on Evolutionary Computation 5(3), pp. 169–191, 2001.
12. Hwang, Y.K., Ahuja, N., A potential field approach to path planning. IEEE Transactions on Robotics and Automation 8(1), pp. 23–32, 1992.
13. Isto, P., Constructing probabilistic roadmaps with powerful local planning and path optimisation. IEEE/RSJ International Conference on Intelligent Robots and System, Lausanne, Switzerland, vol. 3, pp. 2323–2328, 2002.
14. Jiang, K., Seneviratne, L.S., Earles, S.W.E., Finding the 3D shortest path with visibility graph and minimum potential energy. IEEE/RSJ International Conference on Intelligent Robots and Systems, Yokohama, Japan, vol. 1, pp. 679–684, 1993.

15. Kamei, K., Ishikawa, M., Determination of the optimal values of parameters in reinforcement learning for mobile robot navigation by a genetic algorithm. International Congress Series 1269, pp. 193–196, 2004.
16. Kavraki, L.E., Svestka, P., Latombe, J.C., Overmars, M.H., Probabilistic roadmaps for path planning in high-dimensional configuration spaces. IEEE Transactions on Robotics and Automation 12(4), pp. 566–580, 1996.
17. Khatib, O., Real-time obstacle avoidance for manipulators and mobile robots. International Journal of Robotics Research 5(1), pp. 90–98, 1986.
18. Kim, J., Pearce, R.A., Amato, N.M., Extracting optimal paths from roadmaps for motion planning. IEEE International Conference on Robotics and Automation, Taipei, Taiwan, vol. 2, pp. 2424–2429, 2003.
19. Kito, T., Ota, J., Katsuki, R., Mizuta, T., Arai, T., Ueyama, T., Nishiyama, T., Smooth path planning by using visibility graph-like method. IEEE International Conference on Robotics and Automation, Taipei, Taiwan, vol. 3, pp. 3770–3775, 2003.
20. Krink, T., Filipic, B., Fogel, G.B., Thomsen, R., Noisy optimization problems - a particular challenge for differential evolution? Proceedings of the 6th IEEE Congress on Evolutionary Computation, Portland, vol. 1, pp. 332–339, 2004.
21. Latombe, J. C., Robot motion planning. Kluwer Academic Publishers, NY, USA, 1991.
22. Li, B., Jiang, W., Optimizing complex functions by chaos search. Cybernetics and Systems 29(4), pp. 409–419, 1998.
23. Lian, K.Y., Liu, P., Synchronization with message embedded for generalized Lorenz chaotic circuits and its error analysis. IEEE Transactions on Circuits and Systems I: Fundamental Theory and Applications 47(9), pp. 1418–1430, 2000.
24. Lingelbach, F. Path planning using probabilistic cell decomposition. IEEE International Conference on Robotics and Automation, New Orleans, LA, vol. 1, pp. 467-472, 2004.
25. Liu, B., Wang, L., Jin, Y.H., Tang, F., Huang, D. X. Improved particle swarm optimization combined with chaos. Chaos, Solitons, and Fractals, 2005.
26. Luo, C.Z., Shao, H.H. Evolutionary algorithms with chaotic mutations. Control Decision 15, pp. 557–560, 2000.
27. Mahkovic, R., Slivnik, T., Constructing the generalized local Voronoi diagram from laser range scanner data. IEEE Transactions on Systems, Man and Cybernetics - Part A 30(6), pp. 710–719, 2000.
28. Maneewarn, T., Rittipravat, P., Sorting objects by multiple robots using Voronoi separation and fuzzy control. IEEE/RSJ International Conference on Intelligent Robots and Systems, Las Vegas, vol. 2, pp. 2043-2048, 2003.
29. May, R., Simple mathematical models with very complicated dynamics, Nature 261, pp. 459–467, 1976.
30. Melchior, P., Orsoni, B., Lavaialle, O., Poty, A., Oustaloup, A., Consideration of obstacle danger level in path planning using A* and fast-marching optimization: comparative study. Signal Processing 83, pp. 2387–2396, 2003.
31. Nojima, Y., Kojima, F., Kubota, N., Trajectory generation for human-friendly behavior of partner robot using fuzzy evaluating interactive genetic algorithm. IEEE International Symposium on Computational Intelligence in Robotics and Automation, Kobe, Japan, vol. 1, pp. 306–311, 2003.

32. Pei, W., Yang, L., He, Z., Identification of dynamical noise levels in chaotic systems and application to cardiac dynamics analysis. International Joint Conference on Neural Networks, vol. 5, Washington, DC, pp. 3680–3684, 1999.
33. Santos, J., Duro, R.J., Becerra, J.A., Crespo, J.L., Bellas, F., Considerations in the application of evolution to the generation of robot controllers. Information Sciences 133, pp. 127–148, 2001.
34. Schweizer, J., Schimming, T., Symbolic dynamics for processing chaotic signal. ii. communication and coding. IEEE Transactions on Circuits and Systems I: Fundamental Theory and Applications 48(11), pp. 1283–1295, 2001.
35. Shengsong, L., Min, W., Zhijian, H., Hybrid algorithm of chaos optimisation and SLP for optimal power flow problems with multimodal characteristic. IEE Proceedings in Generation, Transmission, and Distribution 150(5), pp. 543–547, 2003.
36. Sierakowski, C. A., Coelho, L. S., Path planning optimization for mobile robots based on bacteria colony approach, 9th Online Conference on Soft Computing in Industrial Applications, Springer-Verlag, London, UK, 2004.
37. Sierakowski, C. A., Coelho, L. S., Study of two swarm intelligence techniques for path planning of mobile robots, 16th IFAC World Congress, Prague, July 4-8, 2005.
38. Sipper, M., On the origin of environments by means of natural selection. AI Magazine 22(4), pp. 133–140, 2002.
39. Storn, R., Differential evolution: a simple and efficient heuristic for global optimization over continuous spaces, Journal of Global Optimization 11(4), pp. 341–359, 1997.
40. Storn, R., Price, K., Differential evolution: a simple and efficient adaptive scheme for global optimization over continuous spaces. Technical Report TR-95-012, International Computer Science Institute, Berkeley, USA, 1995.
41. Takahashi, O., Schilling, R.J., Motion planning in a plane using generalized Voronoi diagrams. IEEE Transactions on Robotics and Automation 5(2), pp. 143–150, 1989.
42. Tsuji, T., Tanaka, Y., Morasso, P.G., Sanguineti, V., Kaneko, M., Bio-mimetic trajectory generation of robots via artificial potential field with time base generator. IEEE Transactions on Systems, Man and Cybernetics - Part C 32(4), pp. 426–439, 2002.
43. Tu, J., Yang, S.X., Genetic algorithm based path planning for a mobile robot. Proceedings of the IEEE International Conference on Robotics Automation, Taipei, Taiwan, pp. pp. 1221–1226, 2003.
44. Watanabe, K., Izumi, K., A survey of robotic control systems constructed by using evolutionary computations. IEEE International Conference on Systems, Man, and Cybernetics, Tokyo, Japan, vol. II, pp. 758–763 1999.
45. Watson, R.A., Ficici, S.G., Pollack, J.B., Embodied evolution: distributing an evolutionary algorithm in a population of robots. Robotics and Autonomous Systems 39, pp. 1–18, 2002.
46. Xiao, J., Michalewicz, Z., Zhang, L., Trojanowski, K. Adaptive evolutionary planner/navigator for mobile robots. IEEE Transactions on Evolutionary Computation 1(1), pp. 18–28, 1997.
47. Yan, X. F., Chen, D.Z., Hu, S.X., Chaos-genetic algorithms for optimizing the operating conditions based on RBF-PLS model. Computers and Chemical Engineering 27, pp. 1393–1404, 2003.

48. Yang, L., Chen, T., Application of chaos in genetic algorithms. Communication Theory Physics 38, pp. 139–172, 2002.

2

Evolving Modular Robots for Rough Terrain Exploration

Olivier Chocron

Laboratoire de recherche en mécatronique,
Ecole nationale d'ingénieurs de Brest,
Technopôle de Brest-Iroise, CS 73822, 29238 Brest Cedex 3,
France
chocron@enib.fr

This chapter proposes an original method for the evolutionary design of robotic systems for locomotion on rough terrain. We encompass the design of wheeled, legged or hybrid robots for their wide range of capabilities for support and propulsion. The goal is to optimize the mechanical and the control system to achieve a locomotion task in a complex environment (irregular, sliding or even with uncertainties). The modular approach brings the possibility to match the diversity of tasks with the combination of assembly modes. This global approach embeds an evolutionary algorithm with a dynamic simulation of the mobile robot operating in its environment. A hybrid encoding of the genotype allows evolving the robot morphology and its behavior simultaneously. Specialized genetic operators have been designed to manipulate this specific encoding and to maintain their efficiency through evolution. Performances are hierarchically evaluated, making decisions based on mechanical analysis and simulation results on line as well as off line. The results are illustrated through a set of design examples that shows how the artificial evolution can in some ways, match human analysis. Some suggestions for hybridizing the method with known techniques and its extension to general complex machine design are given in conclusion.

2.1 Introduction

The use of walking robots have been proposed for locomotion over natural terrains in planetary exploration or in military applications since a score of years [47][31]. What we have learned from the different design experiences of such complex systems is that a dynamic adaptation of the structure and its behavior is more than useful. The main reason for this need of adaptation

O. Chocron: *Evolving Modular Robots for Rough Terrain Exploration*, Studies in Computational Intelligence (SCI) **50**, 23–46 (2007)
www.springerlink.com

is the fact that the mechanical solutions for locomotion on rough terrain are multiples and none are ideal for all situations. Because of slopes, cracks, rocks or variations in the properties of soil in term of granularity, hardness, plasticity or friction, the best strategy to progress will not be constant throughout the mission. These difficulties make of the adaptation issue a global and complex design problem that have been solved diversely. Methodologies for robot design can be split in three classes:

- Pure Engineering
- Knowledge Based Design
- Global or partial Optimization

In the first method, solutions are suggested by accumulated experience of experts. This is the most classical way used in industry but limited to simple systems and difficult to apply in problems where the interactions between the systems and their environment are not a priori known [41]. The second method tries to iteratively improve known solutions accordingly to design heuristics and a data base of physical effects. The last approach starts from scratch and produce incremental improvements by using an objective function and an optimization method such as artificial evolution [24]. It has been proposed for particular machines since a decade or so [50][43].

Many research works have focused on robots adaptation using classical control architectures [30][45], behavior-based [4][1] or evolutionary robotics [10][35][49][25]. Nevertheless few of them have concerned the robot mechanical (i.e. morphology) and control system (i.e behavior) simultaneously [46][32][26], while is it now considered a necessary principle to design embodied intelligence through adaptation [42].

In this chapter, is presented a solution to obtain both mechanical and control design of modular robotic systems (MRS). We extend our previous work on evolutionary robotics for modular manipulators [8][44] and mobile robots [9][21] proposed in [7]. We evolve modular mobile robots with general morphology, with either wheels, legs or both as shown in Fig.2.1. The design principle is a global evolutionary optimization of both their mechanical and control systems in interaction with its environment, based on a dynamic simulation combining the robot and its locomotion task.

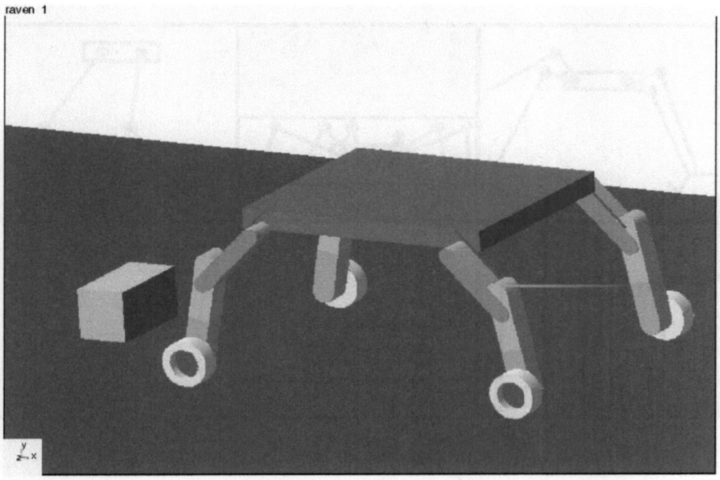

Fig. 2.1. Example of hybrid robot kinematics (ADAMS/View)

In the following sections, we will first introduce the means and goals of this methodology. Then, we describe the different components of the integrated design process proposed. The simulation results for several simple tasks are given and analyzed with regard to the overall approach. Some perspectives for this original method are discussed before drawing a general conclusion.

2.2 Means and Goals

2.2.1 Modular Robotic Systems

To answer the difficult problem of structural adaptation, a promising solution consists in using Modular Robotic Systems (MRS) which are constituted by an assembly of identical mechanical modules [50]. This allows, by reconfiguration, to obtain a large diversity of mechanical structures to match the diversity of robotic tasks as illustrated for locomotion systems in Fig.2.2. Many modular robots have been proposed [27][36] and successfully experimented [20] [28]. The last generation of such robots are able to perform automatic reconfiguration as the M-TRAN from AIST [11] as well as fairly rough terrain locomotion as the famous PolyPod and PolyBot from PARC [13]. What is still to obtain with MRS is the self-transformation initiative and locomotion learning based on task perception.

The main difficulty with the design of these modular systems is to find the optimal module combination to use for a given mission in a given environment. Some good investigations have been made in this objective which proposed algorithms for analysis of modular morphologies [6] or their evaluation [14]. Still, it is not obvious today how a modular robot can adopt the

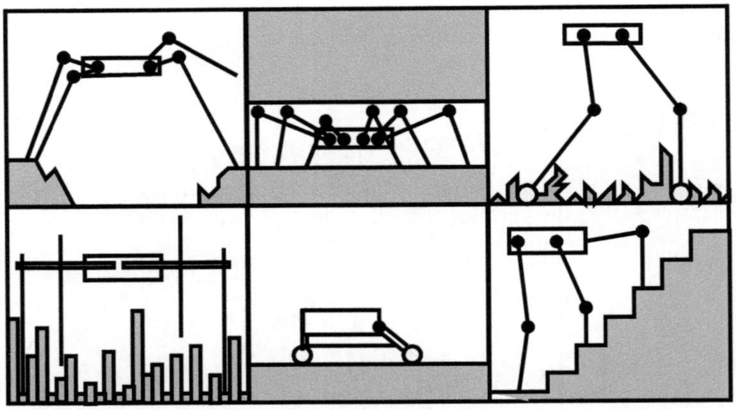

Fig. 2.2. Modular robot in several configurations

right assembly with regard to the task undertaken. Meanwhile, some recent works based on artificial evolution have proved there are solutions to design complex morphologies using generative representations [23] or to match high-level tasks using realistic 3D virtual agents [26]. Stochastic self-assembly can even be used for reconfiguration of modular systems [48]

Another complex issue is the control of such generic systems [19]. Some research works on particular modular robots demonstrated that technological solutions existed to control, connect and power such distributed and independent modules [5][12].

2.2.2 Evolutionary Optimization

Evolutionary Algorithms or Evolutionary Computations are optimization methods based on natural evolution principles [16][22]. The general process of artificial evolution shown in fig.2.3 have been developed for general optimization [18][34], programming [29], engineering design [40], machine learning [15] and robotics [10][38].

The candidate solutions of a given optimization problem are considered as individuals in a population of solutions. The whole population is then evolved with simplified genetic laws and undergoes genetic operators such as **Selection**, **Mutation** and **Crossover**. The mating pool is selected on evaluation criteria (which are gathered in an objective function) and used to produce the next generation of solutions through the genetic mixing and variations. In that way, a population is bred under the *survival of the fittest* law and will evolve toward fitter and fitter individuals and finally, better solutions.

The use of evolutionary computation for modular robot design is easily defensible. To be appropriate an evolutionary algorithm needs a difficult optimization problem and a relevant objective function. Finding the right

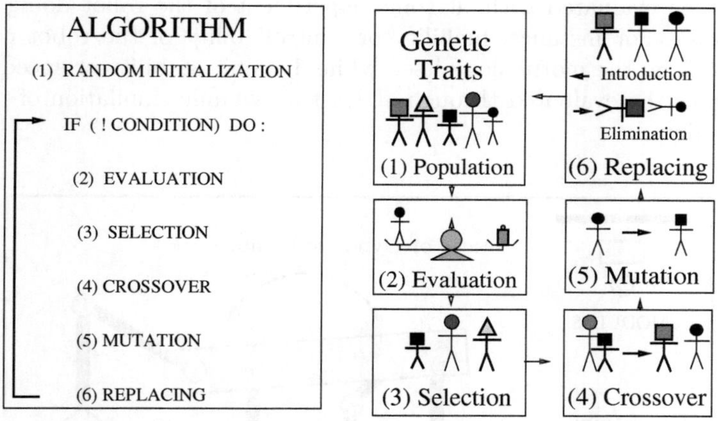

Fig. 2.3. Evolutionary process over a population

assembly among the possibilities of combination is known as combinatorial optimization problem under constraints [39]. This problem belongs to the NP-complete class and thus, has no known algorithm solving it in a finite time [17]. Meanwhile, this kind of problem is not rare in system engineering. It is why many efforts have been made to solve these problems using evolutionary computation [42]. Finding a relevant fitness function for mobile robotics can only be addressed today by experimentation, either in real or computer world [25].

2.2.3 Dynamic Simulation

Evaluating mobile robots on rough terrain is a very difficult task because of the complexity of these systems and their behavior. Describing walking machines for instance is not as fastidious as assessing the effect of interactions between mechanics, control and physical environment. Some mission-based criteria must be precisely defined to evaluate the robot. Figure 2.4 shows some instances of these criteria for a rough terrain exploration task. Optimization criteria that constitute the evaluation process can be split in four:

- Consistency in term of mechanisms or controllability
- Objective to reach in term of mission main goal
- Constraints to be respected during the task
- Performances achieved while doing the task

Some criteria can be determined before experimenting the robot (consistency) while others demand a precise behavioral model to be assessed during the task (constraints or objectives) or after (performances). It appears that most of these criteria depends on the precise activity of the robot and thus, may not be evaluated without some experiment of the robot doing the assigned task. For instance, mobility or controllability of the robot could be analyzed from the morphology itself while dynamic stability or speed performance must be evaluated through at least a dynamic simulation of the task being achieved.

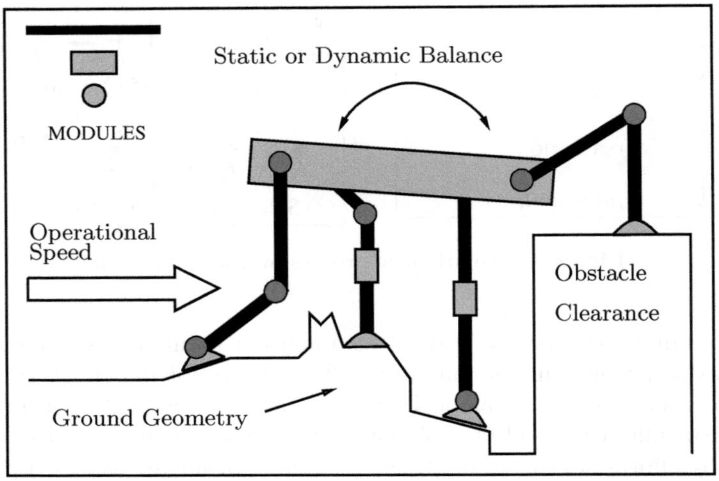

Fig. 2.4. Optimization criteria of walking robots

A mobile robot can be constituted by serial, tree-like or parallel mechanisms and contains kinematic chains that can be closed or opened dynamically according to the robot motions (lifting or putting down a leg). The relative movement of the robot on the ground depends deeply on the interactions between some of its parts (legs, wheels, body) and the ground (free fly, point or surface contact, sliding) and not only on its internal variables. These interactions involve too many parameters and transient causality links to be synthesized in a closed-form solution. The alternative in then clear: Simplify the systems or the evaluation.

The first method is well adapted when we know what kind of solutions we want and then, narrows considerably the domain of the search, generally down to the fairly simple problem of parametric optimization. While this engineering method for morphology is usually preferred, we have chosen the second one because the objective of this work is to obtain new morphological solutions to address the issue of adaptation. For such robots, it is not easy to split up the main locomotion task in subtasks without making a definitive choice on the locomotion strategy and then, to consider several independent stages of evaluation. Another problem is that evolutionary optimization of modular systems can lead us to explore exotic solutions including very variable modes of locomotion (rolling, walking, jumping, crawling,...). It is obvious that it is not feasible to build a mathematical model for each specific kind of assembly [6].

It is why the experimentation by dynamic simulation is used since it provides an objective evaluation of the robot with regard to its task (as the robot whereabouts and states can be directly given by the simulator at any time). Because of the simulation cost in time and the number of times the evaluation must be done, a graphically simplified and numerically approximated simulation has been developed [7]. Other works have made use of virtual environments [26] or simulator co-evolution [3] to allow for realistic and yet time efficient simulations. This tool has been associated to other tools to check consistency, constraints or performances for a hierarchical evaluation of the robot with regard to its task as it is explained in the evaluation (see section 2.3).

2.3 Evolutionary Task-Based Design

2.3.1 Genotype Encoding : Incidence Matrix

The modular kinematics we use to build the robots requires to define mechatronic modules as well as their assembly modes. In addition to payload called the **Base** we use three kinds of modules, the **Segments**, the **Wheels** and the **Joints** (Fig.2.5).

The joints can link the base with segments or wheels and the segments to other segments or wheels. All basic modules of the same kind are strictly identical in dimension, weight and assembly modes. The segments are solids modeled as cylinders and wheels as solid spheres for simplicity of simulation. The revolute joints and represented by two short concentric cylinders including DC-motors and gears.

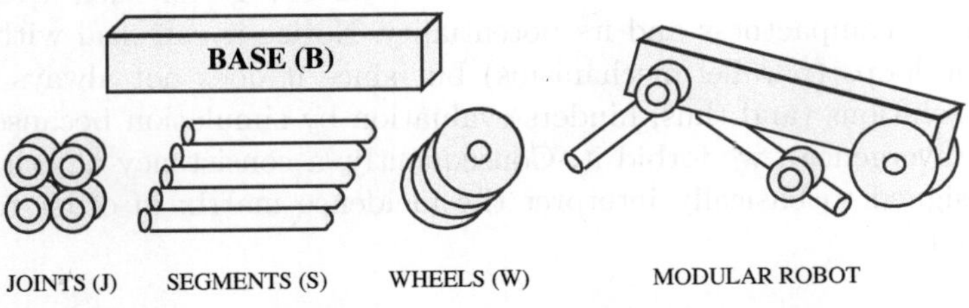

Fig. 2.5. Modular kinematics and assembly example

The resulting modular topology (i.e. morphology) is based on the interactions between the modules. Its matrix representation involves every available modules whether they belong to the topology or not (fixed representation). Their interactions are encoded using an **Incidence Matrix** M in which the solid bodies S_i are represented by the rows and the joints J_j by the columns (see Fig.2.6).

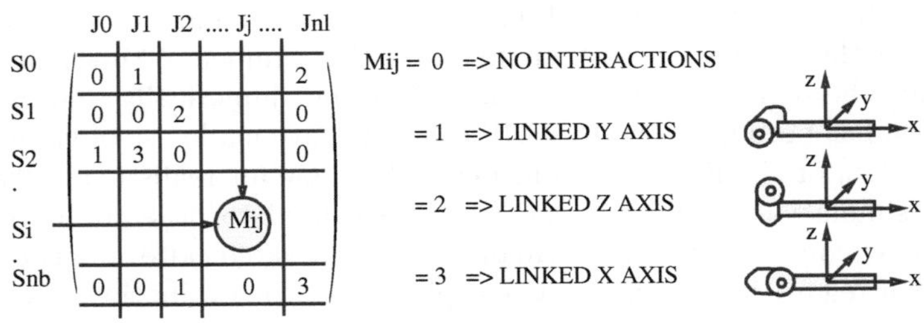

Fig. 2.6. Incidence matrix and interpretation

The integer value M_{ij} at the matrix nodes determines the nature of the linkage between solid S_i and joint J_j. A zero means that there is no linkage between solid and joint and another integer value defines the mode of assembly. The integer value gives the axis (on solid frame) the revolute joint is aligned with (1 for Y, 2 for Z and 3 for X). The base is represented by the first row and the position of attachment points for the joints are distributed on the platform sides at constant intervals. The segments are attached by both extremities (bivalent) and the wheels at their center (univalent). Since the joints and the segments are bivalent, only two elements in the associated column or row can be non zero, only one for the wheels and as many as the number of joints for the base (allowing as many body appendices). This representation is a straightforward and coherent way to encompass a modular assembly [33].

In such a way, we can describe any topological configuration for a given set of modules, whatever their number. The advantage for such a representation is its compactness and its potentiality. Notice it can deal with closed kinematic loops (parallel mechanisms) but since it does not always lead to feasible solutions (and thus, hinders evaluation by simulation because of numerical divergence), we forbid it. Consequently, a consistency algorithm has been designed to basically interpret the incidence matrix in order to reject closed loops.

Enabling dynamic simulation of mobile robot involves designing and applying a control systems, which means control laws. The control laws applied to the joints have to be task-based and dependent of the modular assembly.

Since the same fundamental problem of modeling a generic modular assembly arises when designing a control system, we propose to include it into the genotype and to let the genetic process search for an adapted command in parallel with the topology.

While it is possible (and interesting) to design a modular control system based on command modules (such as PID or non linear controllers) and sensor inputs (position, speed or torques), it is not what is proposed here. The main reason is that the objective of this work is to show the adaptivity of genetically and globally designed systems. This results that if the evolution works, it will search (and eventually find if it exists) a good global solution topology-command however the problem is stated.

We have defined the control law as an open-loop voltage law applied to the actuators (DC-motors and gears) associated to the joint.

The voltage law is defined as follow:

$$U_j = U_{max} * cos(\omega_j * t + \phi_j) \tag{2.1}$$

where
U_j : Voltage applied on joint j
U_{max}: Maximum voltage applicable
ω_j : Pulse of signal j
ϕ_j : Phase of signal j
t : Simulated time

The resulting genotype is constituted by the integer incidence matrix and three float vectors (Fig.2.7). The number N of possible distinct genotype for the structure is then defined as follow:

$$N = (\sum_{k=0}^{k=2} 3^k * C_{Ns}^k)^{Nj} \tag{2.2}$$

where
Ns: Number of solid bodies
Nj: Number of actuated joints

We should multiply this number by R^{Nj} (R being the number of distinct real numbers depending on the used machine) to obtain the total number of possible global genotypes. This number grows exponentially with the number of bodies and joints (see Fig.2.7). This encoding is not canonical for known evolutionary algorithms described in [2], so it forces the design of a new one with its adapted genetic operators.

2.3.2 Topology Operators

For selection, we use the remainder stochastic sampling with replacement [18]. For mutation and crossover, we have designed adapted operators inspired from canonical ones [2].

Matrix Crossover

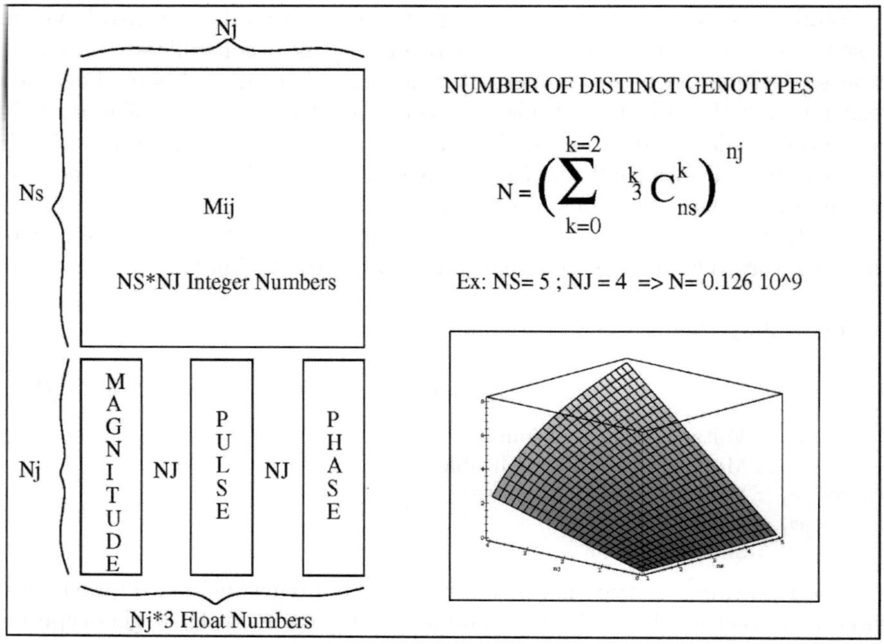

Fig. 2.7. Mixed genotype for a walking robot

The incidence matrix can be considered as a long binary word divided in several rows. So, a classical binary crossover (one point, multi-point or uniform crossover) can be applied (fig2.8).

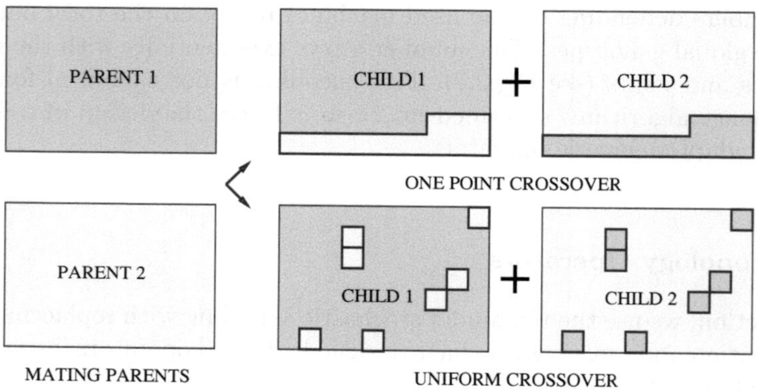

Fig. 2.8. General crossover for a matrix

The consequence is that these operators do not take into account the specificity of the encoded solution and what it does physically represent, (i.e. the modular topology). Hence, two specific crossovers which use problem-based knowledge have been designed to overcome this problem (Fig.2.9). The first crossover is called **Body Crossover**. It exchanges some rows between two matrices in such a way that the associated body linkages are exchanged between two parent robots with the probability X_s. The **Joint Crossover** operates the same kind of exchange on columns and so, between joint linkages with the probability X_j. With these crossovers, the algorithm can exchange structured informations between two parents by selecting for example the leg distribution on the base from one robot and a multi-segment leg from another one.

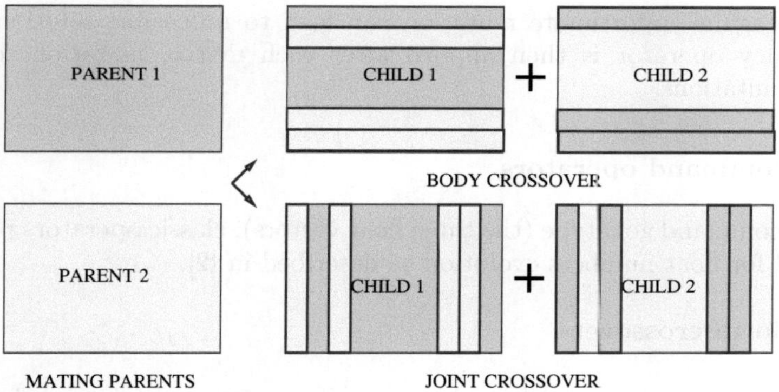

Fig. 2.9. Problem based crossovers

These operators may involve some anomalies in the genotype. These anomalies are of two kinds: structural or geometrical. The first kind includes linkage violations (more than maximum allowed number of linkage) and are easy to detect. The geometrical anomalies are much more subtle and difficult to extract from the incidence matrix. These anomalies include geometrically impossible kinematic closed loops (well known problem in parallel robots) but also separated assemblies from the base and need to be analyzed. Since this kind of genotypes are not valid solutions, we apply the consistency operator after crossover which undoes the last invalid modifications (limiting linkages and eliminating closed loops).

Integer Mutation

For the incidence matrix, we use an integer mutation inspired from binary mutation [18]. The mutation allows each discrete number to mutate into any other discrete value (see Fig.2.10). If the number is zero (no linkage), it can be mutated into a non-zero value (from 1 to 3) and thus, a linkage is created. If the linkage exists, it can be disrupted or re-oriented to a different direction.

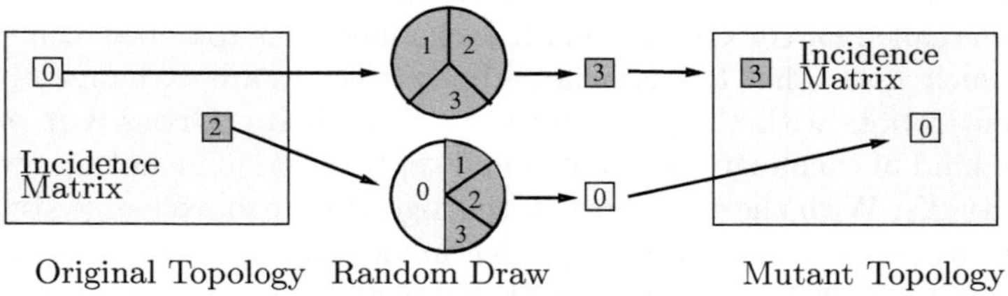

Fig. 2.10. Integer mutation for the incidence matrix

Here again, unfortunate mutation can lead to unfeasible solutions. The consistency operator is then applied after each matrix mutation to undo invalid mutations.

2.3.3 Command operators

For the command genotype (the three float vectors), classic operators patterns are used for float numbers evolution as described in [2].

Uniform crossover

The crossover, inspired from binary uniform crossover gives each pair of float numbers a 50% exchange probability (segregation) and the mutation is directly borrowed to evolution strategy [2].

Here, a random normal number with mean zero and standard deviation σ is added to each float number (Gaussian law, Fig.2.11).

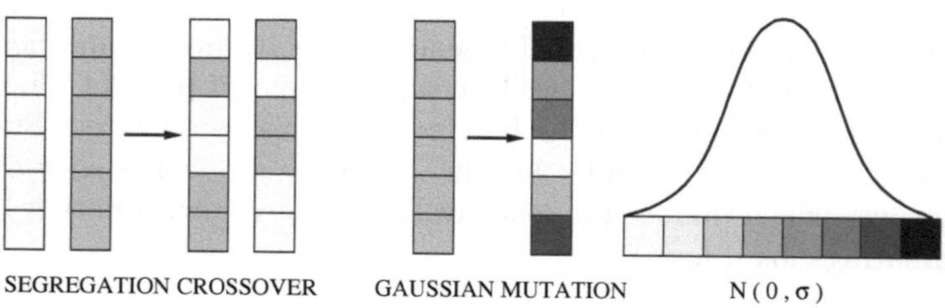

Fig. 2.11. Genetic operators for command vectors

Adaptive mutation

We have applied some adaptability capacities presented in [8] to the mutation operators by adapting the mutation probability for discrete mutation (eq.2.3) and continuous mutation (eq.2.4). The adaptation is based on the relative fitness of the individual.

For highly adapted solutions, the probability is decreased and for unadapted solutions, the probability is increased. In such a way, above average solutions will tend to be kept unmodified from one generation to another while weak ones will likely to be discarded.

- Integer Mutation

$$p_i = \frac{pm_o}{f_i^{\lambda}} \; , \; \lambda = \frac{ln(\frac{p_o}{p_m})}{ln(\tilde{f})} \; , \; p_i \leq 1 \tag{2.3}$$

- Float Mutation

$$\sigma_i = \frac{\sigma_o}{f_i^{\lambda}} \; , \; \lambda = \frac{ln(\frac{\sigma_o}{\sigma_m})}{ln(\tilde{f})} \; , \; \sigma_i \leq 1 \tag{2.4}$$

p_o : Minimum binary mutation probability
σ_o : Minimum mutation standard deviation
p_m: Mean mutation probability
σ_m: Mean mutation standard deviation
f_i : Candidate fitness value
$\tilde{f}$: Population mean fitness

Our previous works have shown the efficiency of this adaptive mutation which allow to keep control on the mutation rate applied for the mean fitness individual (p_m and σ_m) and over the mutation rate applied to to the best possible individual (p_o and σ_o). Notice that the worst case (null fitness) is purely random search.

2.3.4 Phenotype Evaluation

We have to deal with two distinct genetic entities; **topology** and **command**. Both are evolved simultaneously in the same global evolutionary algorithm (Fig.2.12). The evaluation process is done by a approximated dynamic simulation of the robot in its real environment. The simulation uses an efficient numerical approximation to solve the dynamics of a multi body system with environmental interactions.

The evolutionary algorithm calls upon the simulation each time it needs to evaluate a robot for completing the specified task. The simulation being the most time consuming stage, it has to be used scarcely. Indeed, using the complete simulation for all the robots can lead to a very long time of evolution. We use a hierarchical evaluation in three stages :

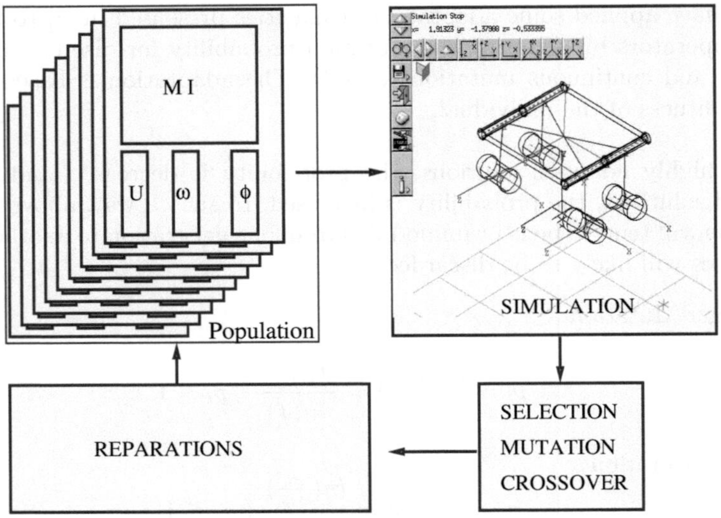

Fig. 2.12. Global evolutionary algorithm

- Mathematical elimination
- Quick simulation
- Full simulation

The mathematical elimination consists in applying the consistency algorithm that checks the validity of linkages and eliminates the separated from base assemblies. The quick simulation is performed for the feasible robots over a short period of time and a moderate accuracy. A performance index in computed during the quick simulation and if it is satisfying, the full simulation is started. The full simulation is done for full time allocation and maximum accuracy. The robot receives a fitness accordingly to its global normalized task performance (see section 2.4).

2.4 Simulation Results

2.4.1 Experimental setup

The experimental setup consists in a computer simulation and the evolutionary algorithm given in sec. 2.3. Evolution and simulation parameters have been designed to obtain a reasonable time and number of generations. Each time, the fitness function is computed relatively to a simple reference solution (engineering design).

The populations have been evolved for four tasks:

- Speeding up
- Joining a goal

- Getting altitude
- Increasing angular momentum

The ground is modeled flat and viscoelastic. This means that reaction forces are proportional to penetration distances and viscous forces to velocities. A model of friction forces has been implemented to take into account solid/solid interactions near zero speed. The robot is constituted by a payload (base) and a variable amount of mechatronic modules as described in table 2.1.

Table 2.1. Mechatronic modules

Solids	**Geometry**	**Mass**	**Friction**	**Nb**
Base	box	40	0.9	1
Legs	cylinder	0.4	0.7	8
Wheels	sphere	4	0.5	8
Joint	**Tmax**	**Wmax**	**Pmax**	**Nb**
Pivot	10 NM	1 rps	10 Hz	8

$Tmax$: Maximum Torque of Joint
$Wmax$: Maximum Angular Speed of Joint
$Pmax$: Maximum Pulse of command signal
Nb : Maximum Number of items

Table 2.2. Simulation parameters

Parameters	**Partial**	**Full**
Simulation Time (s)	1	3
Increment Time (s)	0.005	0.001
Error Max (%)	0.1	0.01

The simulation parameters are given in table 2.2. The genetic operators use parameters of table 2.3. Each evolution run lasts between 6 and 8 hours on a SUN/ULTRA5 workstation.

Table 2.3. Evolution parameters

Population	20	**Generations**	50
p_o	1e-6	p_m	0.01
σ_o	1e-5	σ_m	0.1
X_s	0.5	X_j	0.5

2.4.2 Task 1: Speed up

Here, we try to maximize the final distance (objective function) from the starting position to the end position relatively the x-axis. We evolved hybrid robots (with legs and wheels). The evolution graphic (Fig.2.13) represents the performance of the best individual in the population over fifty generations. It shows that the best robot increases its fitness (globally) and that the whole population follow its lead.

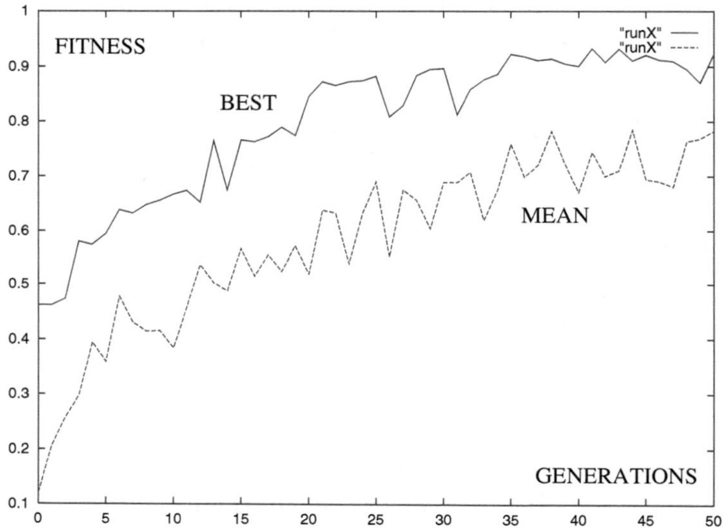

Fig. 2.13. Evolution graph for speed task

The best evolved robot (fig2.14, left) is a wheeled robot, which is not surprising if we consider that wheeled vehicles are the best solution to move fast on flat ground. We can observe that the wheeled robot uses only two wheels it drives to full power (Fig. 2.14) The reason is that the joint torques are large enough to allow the actuators to reach their maximum speed a long time before the simulation end.

It means that additional acceleration is not very useful here and consequently, more wheels would damage the fitness value instead of increase it. The best command evolved simultaneously consists in a high valued input shaped in a flattened parabola (Fig.2.15, right), which maximizes applied torque and provided work to the correctly oriented wheels (pulling along x-axis).

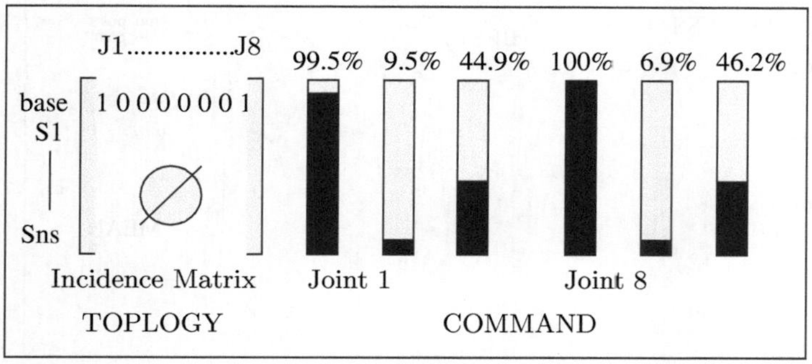

Fig. 2.14. Mixed genotype instance for task 1

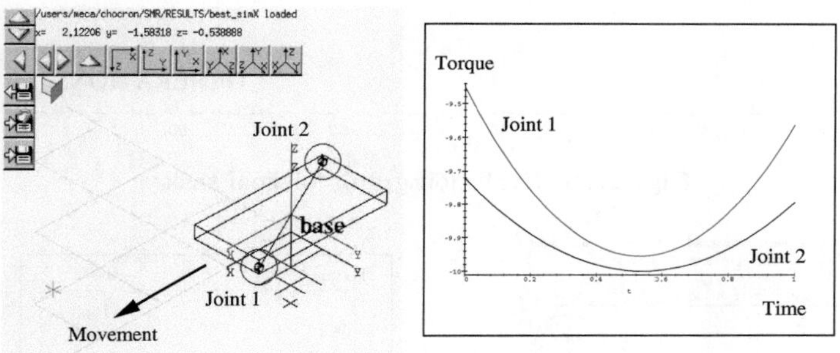

Fig. 2.15. Best robot and best command

2.4.3 Task 2: Reaching a goal

Here, the robot is assigned a goal on the ground to reach at the end of the simulation. The final distance between robot position and desired position is the fitness. The evolution is shown in Fig.2.16.

The evolution this time converges toward a three wheeled robot which rolls toward the desired position by using differential angular wheel speed. Since the linear speed is not useful here, the evolution has accepted an additional wheel to adjust the direction (Fig.2.17, left). The command evolved has favored a highly pulling front wheel and two rear wheels .

2.4.4 Task 3: Getting altitude

Here, we ask the robot to find a way to raise its center of gravity whatever the method. The final altitude of the robot is the fitness. Since it is a hard task to accomplish with wheeled robots, we expect an interesting evolution of legged robots. The evolution graph shows here a rather difficult and long

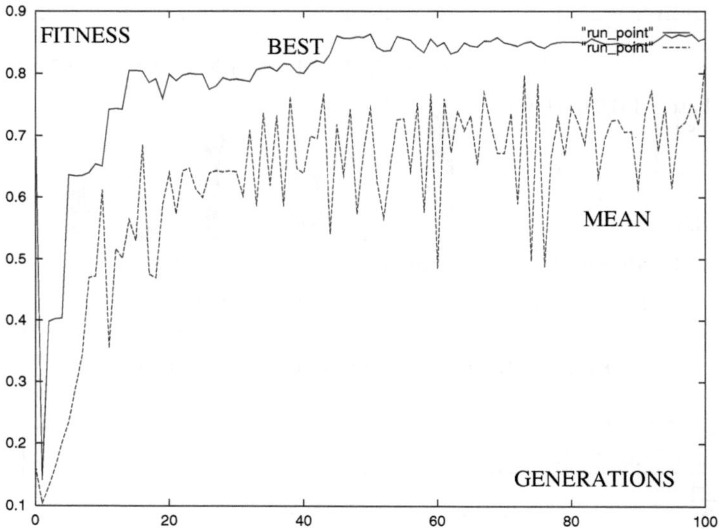

Fig. 2.16. Evolution graph for goal task

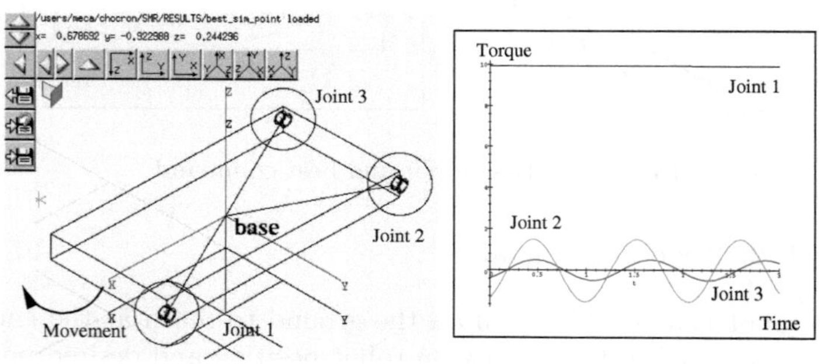

Fig. 2.17. Best robot and best command

convergence(Fig.2.18).

The evolution converges toward a six legged robot (Fig.2.19). We observe than the best ever altitude reached is about 0.5 meters, that is the leg length. Indeed, the best solution our evolution have found to raise its center of gravity over a long period is to give rapid kicks on the ground with a maximum number of legs. This allow the robot to stay on a leg-length altitude for eternity. The fact that not more than one-segment legs are evolved is that multi-segment legs have a far more important inertia, which is not compatible with rapid kicking considering the limited applicable torque.

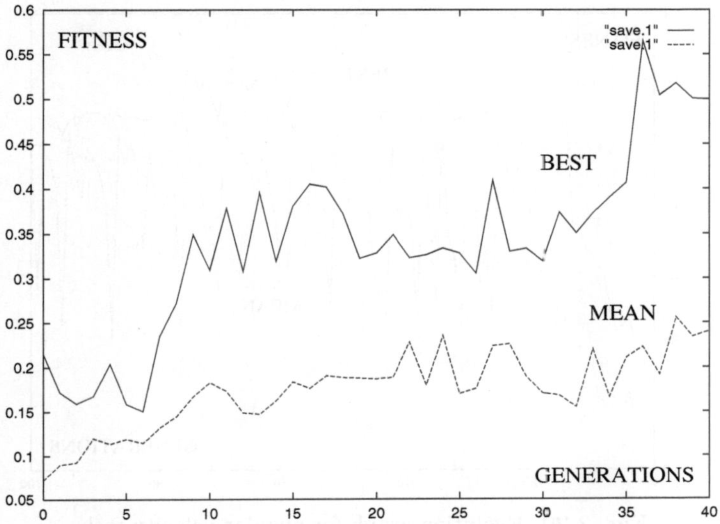

Fig. 2.18. Evolution graph for altitude task

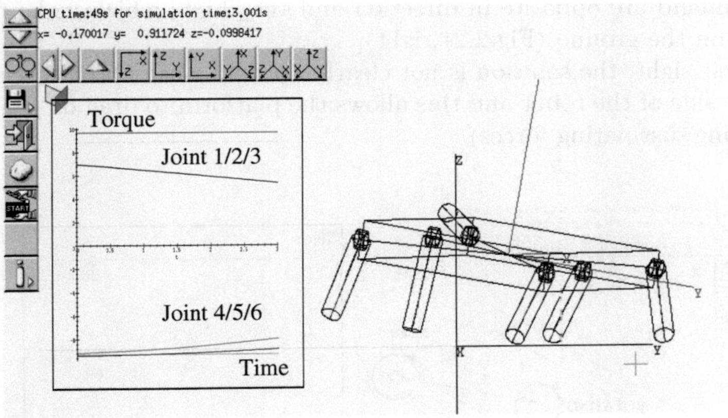

Fig. 2.19. Best robot and best command

2.4.5 Task 4: Increasing angular velocity

Finally, we want to robot to spin quickly along its z axis whatever its position. The fitness is the final angular speed. This kind of task is designed to allow the search for other geometrical orientations of the joint axis and its associated command.

A good solution is found rather quickly but the population have some difficulties to converge (Fig.2.20). The evolved solution is a two-wheeled robot with the wheels on the same side and oriented along y axis (Fig.2.21,left).

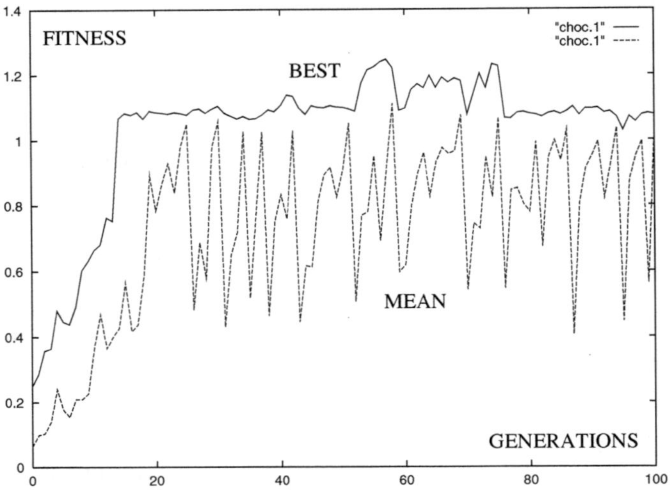

Fig. 2.20. Evolution graph for angular velocity task

The command are opposite in direction and very high, which make the base rotating on the ground (Fig.2.21,right).

At first sight, the solution is not clearly optimal since both wheels are on the same side of the robot and this allows the platform to drag on the ground (generating dissipating forces).

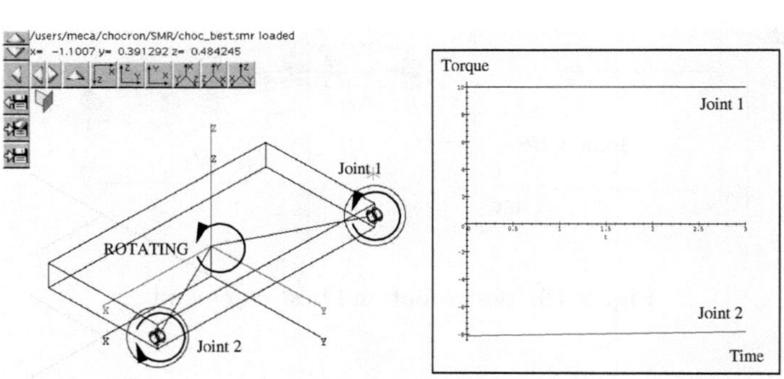

Fig. 2.21. Best robot and best command

But when we consider that the conditions of experiment (torques and mass distribution) and we look at the simulation, it appears that it is a good choice since the body is clearly held above ground during most of the time, eliminating contact forces between body and ground. Two wheels in opposite corners (as we designed for the reference solution) clearly yields less better results as the kinematic lever between traction forces in lesser (suboptimal solution).

From this point of view, the machine has bested the human intuition (if not intelligence). This shows clearly the potential of this approach in engineering complex machines.

2.5 Summary and Conclusions

We have implemented an Evolutionary Algorithm in order to evolve modular robotic systems for various locomotion tasks. The methodology itself is representation and problem independent, adaptable to any robotic design as long as relevant design parameters and a good evaluation process are available. The representation is generic enough to obtain many kinematic topologies and so, to cover a large diversity of robot structures and can be easily improved (allowing new modules or new assembly modes). The adaptations of the encoding, genetic operators and evaluation to the design process has yield valuable results which show some reliability in the algorithm implemented.

Although these experiments are simple and the results not exceptional (8 hours on an SUN ULTRA5 Workstation), they bring some insight and hope in the complex field of evolutionary design the robotic community explored [46].

This work allows to think that robot design can be integrated in an automatic process despite of the complexity of global task modeling by using dynamic simulation.

The primary result of this research is undoubtedly the fact that non explicit fundamentals of the problems can be ' "discovered" ' by the genetic operators.

The secondary result is that the adaptation on genetically evolved robots is such that is can overcome (in some extent) an ill-designed command structure.

To continue this work, we propose to use more efficient behavior adaptations (such as neural networks) while keeping the topology optimization under evolutionary process.

Since morphology and control system are two related subsystems one can co-evolve both entities to find some stable evolutionary trade-off between their adaptation [37]. Our way was to embed both sub-systems in the same global genotype as it is in natural creatures and as we continued using artificial neural networks [21].

Moreover, the issue of evolution time has compelled us to find some ways to use a simulation for evaluation. While our simulation does not match the visual rendering and the interactions diversity of virtual environments as proposed in [46][26], it is realistic and explicit enough to grasp the behavior of 3D modular robots.

In a near future, it seems possible to bring the evaluation in the real world using automatic design and manufacture of robots as done in [32].

References

1. R.C. Arkin. *Behavior-Based Robotics.* MIT Press, Cambridge (MA,USA), 2000.
2. T. Bäck. *Evolutionary Algorithms in Theory and Practise.* Oxford University Press, New York, 1996.
3. J. Bongard and H. Lipson. Once more unto the breach: Co-evolving a robot and its simulator. In *Proceedings of the Ninth International Conference on the Simulation and Synthesis of Living Systems (ALIFE9)*, pages 57–62, 2004.
4. R.A. Brooks. *Cambrian Intelligence.* MIT Press, Cambridge (MA, USA), 1999.
5. Z. Butler, R. Fitch, and D. Rus. Distributed control for unit-compressible robots: Goal-recognition, locomotion and splitting. *IEEE/ASME Transactions on Mechatronics*, 7(4):418–430, 2002.
6. I. M. Chen and J. Burdick. Determining task optimal modular robot assembly configurations. In *IEEE International Conference on Robotics and Automation (ICRA)*, Minneapolis (MN, USA), 1996.
7. O. Chocron. *Conception évolutionnaire de Systèmes Robotiques.* Thèse de doctorat, Université Pierre et Marie Curie (Paris 6), 2000.
8. O. Chocron and Ph. Bidaud. Evolutionnary algorithms in kinematic design of robotic systems. In *IEEE/RSJ International Conference on Intelligent Robots and Systems (IROS)*, Grenoble (France), October 1997.
9. O. Chocron and Ph. Bidaud. Evolutionary algorithm for global design of locomotion systems. In *IEEE/RSJ International Conference on Intelligent Robots and Systems (IROS)*, Kyongju (South Korea), October 1999.
10. Y. Davidor. *Genetic Algorithms and Robotics: A Heuristic Strategy for Optimization.* World Scientific, 1991.
11. E. Yoshida et al. A self-reconfigurable modular robot : Reconfiguration planning and experiments. *International Journal of Robotic Research*, 21(10), October 2002.
12. J. Brufau et al. Micron: Small autonomous robot for cell manipulation applications. In *IEEE International Conference on Robotics and Automation (ICRA)*, Barcelona, 2005.
13. M. Yim et al. Modular reconfigurable robots in space applications. *Journal of Autonomous Robot*, 14(2):225–237, March 2003.
14. S. Farritor and S. Dubowsky. On modular design of field robotic systems. *Journal of Autonomous Robots*, 10(1):57–65, 2001.
15. D.B. Fogel. *Evolutionary Computation. Toward a New Philosophy of Machine Intelligence.* IEEE Press, 1995.
16. L. J. Fogel, A. J. Owens, and M. J. Walsh. *Artificial intelligence through simulated evolution.* New York:Jhon Wiley, 1966.
17. M.R. Garey and D.S. Johnson. *Computers and Intractability : A Guide to the Theory of NP-Completeness.* W.H. Freeman and Co., 1979.
18. D.E. Goldberg. *GENETIC ALGORITHMS in Search, Optimization and Machine Learning.* Addison Wesley, 1989.
19. B. Goodwine and J. Burdick. Gait controllability for legged robots. In *IEEE International Conference on Robotics and Automation (ICRA)*, Leuven (Belgium), 1998.
20. G.J. Hamlin and A.C. Sanderson. Tetrobot: A modular approach to parallel robotics. *IEEE Robotics and Automation Magazine*, 4(1):42–50, March 1997.

21. P. Henaff and O. Chocron. Adaptive learning control in evolutionary design of mobile robots. In *IEEE International Conference on Systems, Man and Cybernetics*, Hammamet (Tunisia), October 2005.
22. J. H. Holland. *Adaptation in natural and artificial systems: An Introductory Analysis with Applications to Biology, Control, and Artificial Intelligence*. Bradford Book, MIT Press, 1975.
23. S. Hornby, H. Lipson, and J.B. Pollack. Generative representations for the automated design of modular physical robots. *IEEE Transactions on Robotics and Automation*, 19(4):703–719, 2003.
24. Ph. Husband J-A. Meyer and I. Harvey. Evolutionary robotics: A survey of applications and problems. In *First European Workshop, EvoRobot98*, pages 1–21, Paris, 1998. Springer.
25. N. Jakobi. Running across the reality gap: Octopod locomotion evolved in a minimal simulation. In *First European Workshop, EvoRobot98*, pages 39–58, Paris, 1998. Springer.
26. M. Komosinski. The framsticks system: versatile simulator of 3d agents and their evolution. *Kybernetes*, 32:156 – 173, February 2003.
27. K. Kotay and D. Rus. The self-reconfiguring robotic molecule. In *IEEE International Conference on Robotics and Automation (ICRA)*, Leuven (Belgium), 1998.
28. K.D. Kotay and D.L. Rus. Task-reconfigurable robots: Navigators and manipulators. In *IEEE/RSJ International Conference on Intelligent Robots and Systems (IROS)*, Grenoble (France), 1997.
29. J. R. Koza. *Genetic Programming II: Automatic Discovery of Reusable Programs*. The MIT Press, 1992.
30. J.P. Laumond. *La robotique mobile*. Hermes, 2001.
31. Christopher Lee and Yangsheng Xu. $(dm)^2$: A modular solution for robotic lunar missions. In *Space Technol.*, volume 16. Elsevier Science Ltd, 1996.
32. H. Lipson and J.B. Pollack. Automatic design and manufacture of robotic lifeforms. *Nature*, 406:974–978, 2000.
33. T. Matsumaru. Design and control of the modular robot systems: Tomms. In *IEEE International Conference on Robotics and Automation (ICRA)*, 1995.
34. Z. Michalewicz. *Genetic Algorithms + Data Structures = Evolution Programs*. Springer-Verlag, 1992.
35. O. Miglino, H.H. Lund, and S. Nolfi. Evolving mobile robots in simulated and real environments. Technical report, University of Palermo, Italy, 1996.
36. S. Murata, H. Kurokawa, E. Yoshida, K. Tomita, and S. Kokaji. A 3-d self reconfigurable structure. In *IEEE International Conference on Robotics and Automation (ICRA)*, Albuquerque (NM, USA), 1997.
37. S. Nolfi and D. Floreano. How co-evolution can enhance the adaptive power of artificial evolution: implications for evolutionary robotics. In *First European Workshop, EvoRobot98*, pages 22–38, Paris, 1998. P. Husbands and A. Meyer, Springer.
38. S. Nolfi and D. Floreano. *Evolutionary Robotics: The Biology, Intelligence, and Technology of Self-Organizing Machines*. Bradford Books, 2000.
39. C.H. Papadimitriou and K. Steiglitz. *Combinatorial Optimization- Algorithms and Complexity*. Dover Publications, 1982.
40. I.C. Parmee. *Evolutionary and Adaptive Computing in Engineering Design*. Springer, 2001.

41. S. Perkins and G. Hayes. Robot shaping - principles, methods and architectures. In *Workshop on Learning robots and Animals at Adaptation in Artificial and Biological Systems Conference (AISB)*, Brighton, 1996.
42. R. Pfeifer, F. Iida, and J. Bongard. New robotics: Design principles for intelligent systems. *Artificial Life, Special Issue on New Robotics*, 11(1-2):99–120, 2005.
43. G.P. Roston. *A Genetic Methodology for Configuration Design*. PhD thesis, Dpt. of Mechanical Engineering, Carnegie Mellon University, 1994.
44. S. Sakka and O. Chocron. Optimal design and configurations of a mobile manipulator using genetic algorithms. In *Proceedings of 10th IEEE International Workshop on Robot and Human Interactive Communication,ROMAN01*, pages 268 – 273, 2001.
45. R. Siegwart and I.R. Nourbakhsh. *Autonomous Mobile Robots*. MIT Press, 2004.
46. K. Sims. Evolving 3d morphology and behavior by competition. In *Artificial Life IV Proc.*, pages 28–39, Cambridge (MA, USA), 1994. MIT Press, ed by R. Brooks and P. Maes.
47. D. J. Todd. *Walking Machines : An Introduction to Legged Robots*. Kogan Page Ltd, London, 1985.
48. P. White, V. Zykov, J. Bongard, and H. Lipson. Three dimensional stochastic reconfiguration of modular robots. In *Proceedings of Robotics: Science and Systems*, pages 161–168. MIT Press, Cambridge, MA, 2005.
49. J. Xiao, Z. Michalewicz, L.Zhang, and K. Trojanowski. Adaptive evolutionary planner/navigator for mobile robots. *IEEE Trans. Evolutionary Computation*, 1(1):18–28, 1997.
50. M. Yim. A reconfigurable modular robot with many modes of locomotion. In *Proceedings of Int. Conf. on Advanced Mechatronics*, Tokyo, 1993. IEEE.

3

Evolutionary Navigation of Autonomous Robots Under Varying Terrain Conditions

Terrence P. Fries

Department of Computer Science, Coastal Carolina University,
Conway, South Carolina 29528 USA
tfries@coastal.edu

Optimal motion planning is critical for the successful operation of an autonomous mobile robot. Many proposed approaches use either fuzzy logic or genetic algorithms (GAs), however, most approaches offer only path planning or only trajectory planning, but not both. In addition, few approaches attempt to address the impact of varying terrain conditions on the optimal path. This chapter presents a fuzzy-genetic approach that provides both path and trajectory planning, and has the advantage of considering diverse terrain conditions when determining the optimal path. The terrain conditions are modeled using fuzzy linguistic variables to allow for the imprecision and uncertainty of the terrain data. Although a number of methods have been proposed using GAs, few are appropriate for a dynamic environment or provide response in real-time. The method proposed in this research is robust, allowing the robot to adapt to dynamic conditions in the environment.

3.1 Introduction

Optimal motion planning is essential to the successful operation of an autonomous mobile robot. For many applications, it is imperative that an autonomous robot be able to determine and follow an optimal of near optimal path to a destination. Motion planning is composed of two functions: path planning, and trajectory planning [1, 2]. Path planning generates a collision-free path through an environment containing obstacles. The path is optimal with respect to some selected criterion. Trajectory planning schedules the movements of the robot along the planned path.

Many approaches to motion planning have been proposed. However, most approaches address only path planning or only trajectory planning, but not both [1, 3, 4, 5]. The GA coding scheme used in this research combines path planning with trajectory planning, thus, eliminating the additional step of

T. P. Fries: *Evolutionary Navigation of Autonomous Robots Under Varying Terrain Conditions*, Studies in Computational Intelligence (SCI) **50**, 47–62 (2007)
www.springerlink.com

trajectory planning once an optimal path is found and reducing the computational time to allow a real-time response.

Implementation issues are a primary consideration when evaluating autonomous robot navigation algorithms. To be of use, a motion planning method must be sufficiently efficient to execute in real-time with the limited onboard computational resources common on autonomous mobile robots.

It is common for GA-based approaches to motion planning to function only in a static environment due to the processing time required to produce an optimal solution [1, 3, 6, 7, 8]. However, many applications require that the robot respond to a changing environment and moving obstacles. In many operational situations, it is impractical to assume that the environment will not undergo changes and that all objects will be stationary. This research provides a method that allows the robot to function in a dynamic environment.

In most cases, GAs do not provide real-time solutions to motion planning problems [1, 3, 6, 7, 8]. Those that do offer real-time response usually have unacceptable restrictions, such as limiting solutions to x-monotone or y-monotone paths [9]. An x-monotone path is one in which the projection of the path on the x-axis is non-decreasing. This places an unacceptable restriction on the solution path because even a simple path between two rooms in a building is neither x-monotone nor y-monotone as shown in Fig. 3.1.

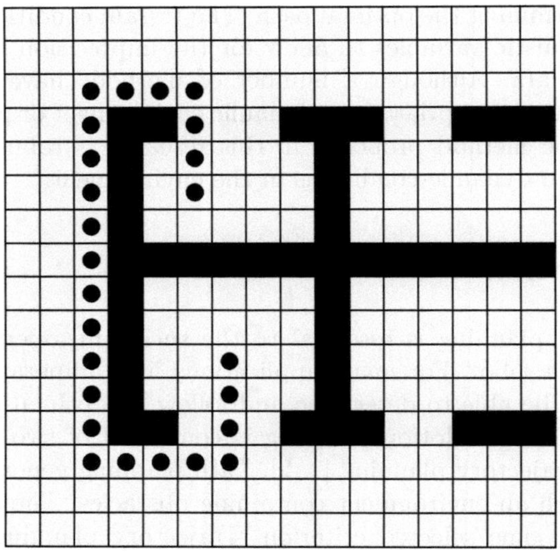

Fig. 3.1. Non-monotone path between rooms

In an effort to reduce the computation time, some researchers have proposed encoding all chromosomes with a fixed length [9, 10]. However, it has been shown that for robot path planning fixed length chromosomes are too restrictive on the solution path by placing unnecessary constraints on the representation of the environment and on the path [6, 10].

Research into using genetic algorithms for path planning include the work of Shibata and Fukuda [11] who proposed a motion planning strategy for a point robot in a static environment. Davidor [12] proposed GA approach attempts to minimize the accumulated deviation between the actual and desired path. However, this assumes that a desired path is already known. Nearchou [13] presented an approach using GAs that compares favorably with other evolutionary techniques, but it requires that the map be converted to a graph. None of these approaches account for dynamic conditions.

A further restriction among current motion planning approaches is that few approaches consider varying terrain conditions with most labeling an area either free of obstacles or totally blocked [8, 10]. In many real world cases, an area may be composed of terrain that is difficult to traverse. Difficult terrain may include sandy areas which cause slippage, rocky areas that require minor course adjustments within them and/or loss of time, or sloped areas that may cause slippage or increased time to climb. Such terrain may be traversable at the cost of increased time, but provide a more optimal path than totally clear terrain.

Iagnemma and Dubowsky [14] are concerned with the roughness of the terrain as characterized by elevation changes. However, the relatively large elevation changes which are addressed lead to consideration of wheel diameter and wheel-terrain contact force and angle. This complicates the computation to such a degree as to preclude a real-time response.

Other researchers [15, 16, 17, 18, 19, 20, 21, 22, 23, 24] also consider issues of severe terrain such as rocky areas and large elevation changes. However, other terrain factors are equally influential on the navigability of a section of terrain with what may appear to be rather benign elevation changes. Features such as a sandy or gravel surface or a mild slope may also impact terrain traversability. Many simple, wheeled robots face these terrain issues even when not operating in rocky environments with extreme elevation changes.

One must also consider the uncertainty inherent in terrain sensing devices and mapping methods [17, 18, 25, 26]. In these cases, the exact nature of the terrain in a particular area is uncertain and, as a result, only an estimate of its true condition. Terrain measurement uncertainty has been addressed by research using methods such as a Gaussian [14] and Kalman [27] filters. These methods are usually computationally intensive and, therefore, inappropriate for real-time operation. The approach presented here addresses the issue of terrain uncertainty by assigning to each section of terrain a fuzzy linguistic variable which indicates the difficulty of traversing that area.

The evolutionary approach to motion planning described in this chapter provides real-time motion planning in a dynamic environment without the restrictions of monotone paths or fixed length chromosomes. It also allows terrain to be labeled with the difficulty of traversal, thus, allowing it to be considered as part of a solution path.

Section 3.2 presents the representation of the environment and GA basics. In Sect. 3.3, the new fuzzy genetic motion planning approach is presented. Section 3.4 provides a discussion of the implementation and test results of the new motion planning approach. Section 3.5 provides a summary of the approach and discusses future directions for research on this method.

3.2 Problem Formulation

Several assumptions have been imposed on the motion planning approach to reduce the initial complexity of the problem and allow concentration on the particular issues of motion planning and terrain considerations. First, the mobile robot is assumed to be holonomic which means it is capable of turning within its own radius. This eliminates the need to consider the complexity in the movement of nonholonomic robots [2, 16]. The robot is limited to the move-stop-turn course of action, where after moving, the robot will stop, turn, and then proceed to move again. This avoids issues related to terrain-induced lateral forces on the wheels while turning in an arc while moving. For purposes of simplification robot localization uncertainty is ignored. Localization uses only dead reckoning based on wheel odometry information. Localization problems are to be addressed in future research.

3.2.1 Environment Grid

The environment in which the robot will maneuver is divided into an environment grid and a path is described as a movement through a series of adjacent cells in the grid. This representation is an extension of the occupancy grid which has been a common method of representing a robot's operational domain [2, 28]. The length of the path $d\ (a,\ b)$ between two adjacent cells a and b is defined as the Euclidean distance between the centers of the two cells. This representation of distance allows the map data to be stored in any efficient format, such as a quadtree [29]. Storage by such methods provides more compact representation of an environment by storing large obstacles as a single grid location, rather than many uniformly sized small squares. It also allows the path to be represented by fewer grid transitions, thus, reducing the size of the GA encoding string, or chromosome, and the time required to determine a solution. Each cell in the grid is assigned a fuzzy value that indicates the difficulty in traversing the terrain in that cell. The use of fuzzy values allows cells with moderately hostile terrain, such as rocks or loose sand, to be considered in a possible solution path while being weighted by their difficulty of traversal. A cell which contains an obstacle is assigned a fuzzy value indicating it is impassable and any possible solution path containing it is unacceptable. For this chapter, the grid will be restricted to 16 by 16 for simplicity, however, the algorithm has been successfully tested for much larger sized grids. Further discussion of this restriction and actual testing is found in the Test Results in Sect. 3.4.

For purposes of this research, the robot is considered to be a holonomic point, that is, it is able to turn within its own radius. Because the robot is holonomic, a path can change direction within a cell and does not require a large arc for turning. Since it is a point, when traversing between two diagonally adjacent cells, it is not necessary to consider the other cells sharing the common corner as shown in Fig. 3.2. This is not as impractical as it may appear at first glance. All real obstacles are expanded by half the radius of the robot when marking which cells are obstructed, thus allowing the robot to be treated as a point. This permits navigation of the center of the robot along the side of an obstacle or diagonally between obstacles. In Fig. 3.2, the actual obstacle is solid and the expansion is shaded.

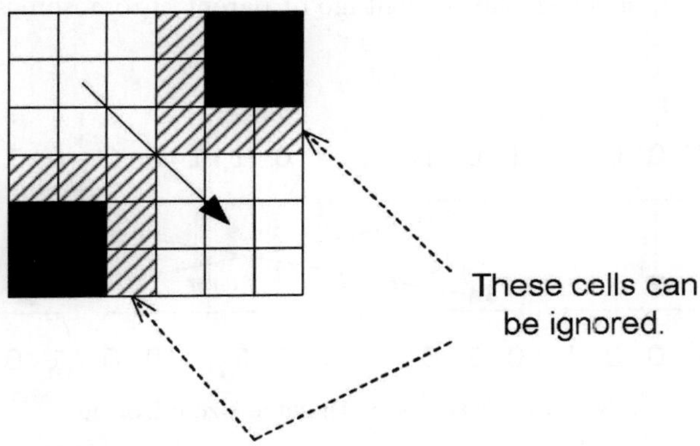

Fig. 3.2. Diagonal traversal of cells

3.2.2 Genetic Algorithms

A genetic algorithm [30, 31] for optimization commonly represents a possible solution as a binary string, called a chromosome. Numerous approaches have been proposed for encoding paths as binary strings.

The GA begins with an initial population of chromosomes, or possible solutions. The GA then creates new individuals using methods analogous to biological evolution. The fitness of each chromosome is calculated using a fitness function. The criteria for evaluation is domain specific information about the relative merit of the chromosome. For example, in the case of path planning, the fitness function may calculate the time required or distance traveled to move from the initial location to the goal. The fittest parents are chosen to reproduce to create offspring. The offspring are generated by subjecting the parent chromosomes to various genetic operators including crossover and mutation.

The crossover operator combines parts of two different chromosomes to create two new ones. In single point crossover, the left part of a chromosome is combined with the right part of another, and then the remaining two parts of the originals are combined, thus, creating two new offspring. This type of crossover produces two offspring of the same size as the parents as shown in Fig. 3.3. The two sections can also be combined to form offspring of differing sizes as shown in Fig. 3.4. The crossover point is usually randomly selected, although it can be fixed for particular applications. Multiple point crossover divides the chromosome into multiple strings which are recombined with those of another chromosome. A fixed number of divisions can be specified for all multiple point crossovers, or the number of partitions can be randomly set for each pair of chromosomes being operated on. Additional crossover schemes which utilize heuristics also exist, but add too much computational complexity for this application. Not all chromosomes are subjected to crossover. The crossover rate, γ, specifies the percentage of parent chromosomes involved in the crossover.

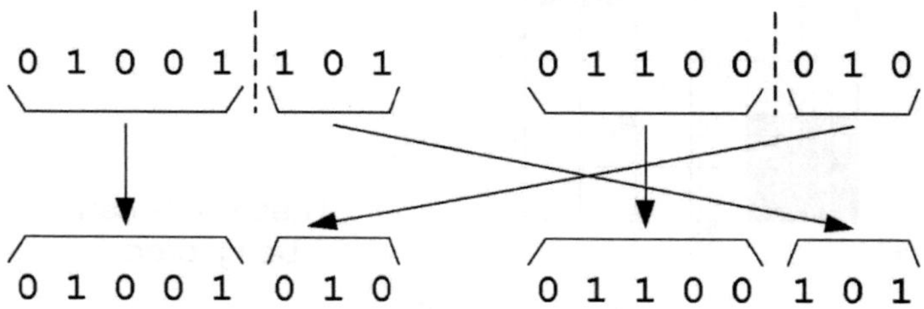

Fig. 3.3. Crossover with same size offspring

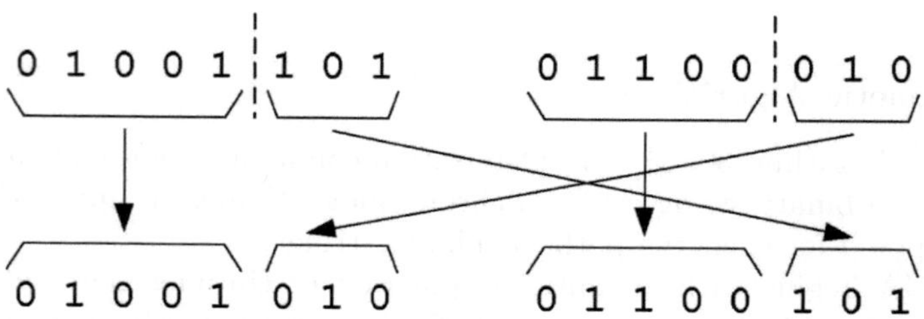

Fig. 3.4. Crossover with different size offspring

3.3 Motion Planning Algorithm

Several components can significantly affect the performance of a genetic algorithm: encoding of the chromosome, initial population, genetic operators and their control parameters, and fitness function.

3.3.1 Encoding the Chromosome

The first step is to choose a coding scheme which maps the path into a binary string or chromosome. Emphasis is placed on minimizing the length of the binary string. Minimizing the length of the chromosome reduces the number of generations necessary to produce an acceptable solution because less permutations are possible. A variable length string composed of blocks which encode the direction of movement and the length of the movement was chosen. Consider the robot in the center cell as in Figure 3.5 (a) having just arrived from cell 4 and facing in the direction of the arrow. There are eight possible directions for movement. However, cell 4 can be eliminated from consideration for the next move since the robot came from that cell and returning to it would create a non-optimal path. Cells 1, 2, 6, and 7 can be eliminated because they could have been reached from cell 4 using a shorter distance than through the center cell in which the robot curre.y is positioned. Only three cells remain in consideration for possible movement. The three cells require only 2 bits to encode as in Figure 3.5 (b).

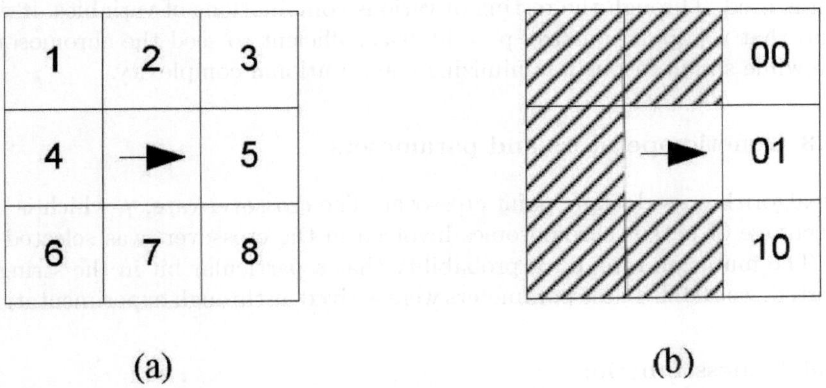

Fig. 3.5. Possible movement to next cell

The largest number of cells that can be traversed in a square grid is found by starting in a corner and moving as far as possible along a side or the diagonal. Since the grid is constrained to 16 by 16 cells, the maximum number of cells that can be traversed in a single move is 15 which requires 4 bits to encode. As a result, each movement can be encoded in a 6-bit block as shown in Fig. 3.6. For larger n x n grids, the block size would be $2 + log_2 n$. A chromosome composed of these 6-bit blocks contains not only the path, but also the necessary trajectory information for movement of the robot. Thus, this unique encoding provides both path planning and trajectory planning.

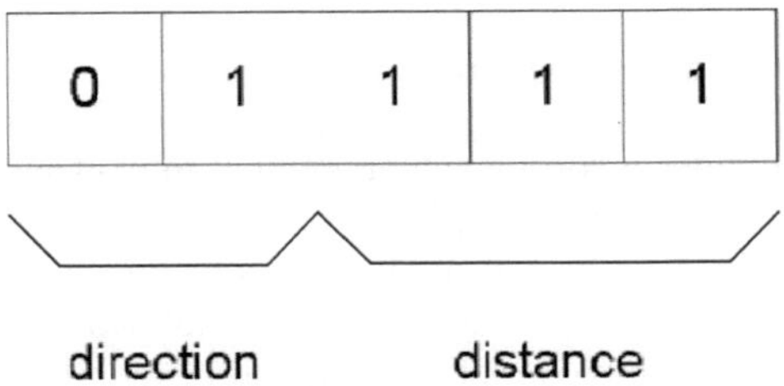

Fig. 3.6. Block encoding of one movement

3.3.2 Initial Population

The motion planning approach begins by randomly generating an initial population of chromosomes. In an effort to direct the solution to the shortest path, another chromosome is added to the initial population. It represents a straight line from the start to destination regardless of obstacles. If a straight line is not possible due to grid locations, the closest approximation to a straight line path is used. Through the testing of various combinations of variables, it was found that a population size, $p = 40$, was sufficient to seed the chromosome base while simultaneously minimizing computational complexity.

3.3.3 Genetic operators and parameters

The algorithm used single point crossover. The crossover rate, γ, which is the percentage of parent chromosomes involved in the crossover, was selected as 0.8. The mutation rate, μ, or probability that a particular bit in the string is inverted, was 0.02. These parameters were arrived at through experimentation.

3.3.4 Fitness function

Selection of a fitness function is a critical aspect of this research. Chromosomes are selected for reproduction through crossover and mutation based on the fitness function. The value provided by the fitness function is then used to retain the best members of the population for the next generation. Common approaches to using GAs for path planning set the fitness to an unacceptable value for any chromosome whose path traverses a grid cell with an obstacle in it. Otherwise, the fitness is based upon the distance traveled in the path.

However, this does not account for terrain conditions. In an effort to consider adverse terrain conditions, each cell is assigned a value corresponding to the difficulty in traversing its terrain. The difficulty in traversing a particular terrain is imprecise because it may vary from one instance to another. In addition, it is problematical to compare different terrain conditions because of the varied nature of each. Further difficulty in a assigning a precise terrain difficulty exists because traversal of an cell in different directions can have significantly different difficulty levels. For example, traversing a sandy hill moving downhill, uphill, or across the side of the hill have dissimilar difficulty levels. Because of the imprecision of terrain conditions and the problems in directly comparing them, this research has chosen to express the terrain difficulty as fuzzy numbers. The terrain condition for each cell is expressed as a triangular fuzzy number using the linguistic variables shown in Fig. 3.7. This uniform distribution of fuzzy linguistic variables is common in fuzzy logic applications. The allotment of linguistic variables over the domain [0, 1] provides equal coverage for each possible terrain condition. Experimental trials with uneven distributions were conducted but they proved to be less effective in the algorithm. Terrain conditions represent the difficulty in traversing the cell which can be affected by conditions such as slope, sand, rocks, etc. As a result, the fitness function must be expanded for this research. For any path not passing through an obstacle, the fitness function uses the Euclidean distance between the centers of the cells traversed weighted by the terrain conditions for each cell.

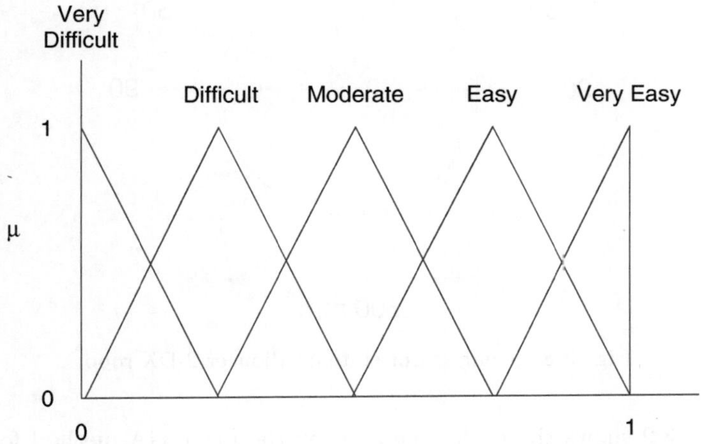

Fig. 3.7. Fuzzy representation of terrain conditions

3.3.5 Dynamic Environment

The fuzzy genetic motion planning method allows the robot to function in a dynamic environment. If an obstacle is detected by the robot where it not expected, the planner simply recalculates a new optimal path in real-time and the robot can continue its movement.

3.4 Test Results

The test software was implemented using C++ and Saphira robot control software by SRI International. It was tested first in the Saphira simulator and, then, on a Pioneer 2-DX mobile robot. The Pioneer 2-DX is a holonomic 3-wheeled robot with a 250 mm radius. It is equipped with a suite of eight sonar sensors arranged as shown in Fig. 3.8 and tactile bumpers. A predefined map representing the environment as a grid was provided to the robot. For clarity in presenting the test results, all results are shown for a 16 by 16 grid. This allows the demonstration of the algorithm's functionality while maintaining readability of the images. Testing has also been conducted using much larger grids and quadtree representations of the environment.

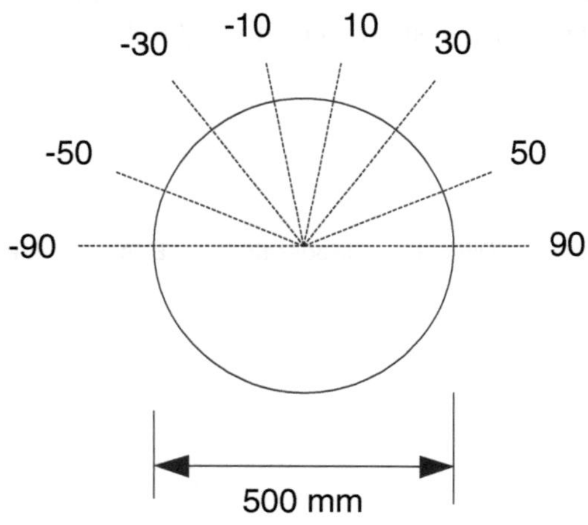

Fig. 3.8. Sonar sensor suite on Pioneer 2-DX robot

Figure 3.9 shows the path generated by the fuzzy GA method for a particular environment with no cells labeled with terrain difficulty values. The S and D indicate the start and destination cells, respectively, of the robot, and the black cells indicate solid obstacles. Manual examination confirms that this is an optimal path. This solution required seven 6-bit blocks in the optimal solution chromosome, including one to turn the robot to a starting orientation before beginning movement.

Next the labeling of terrain difficulty with fuzzy values was verified. The shaded cells on the grid in Fig. 3.10 were labeled as having Moderate difficulty to traverse. This had no effect on the generation of an optimal path, as should be the case. The path generated was the same as in the first test without difficulty labeling. However, when the labeling of the same area of difficulty was changed to Difficult, a different path was produced by the fitness function as shown in Fig. 3.11. When the Moderate area was enlarged as in Fig. 3.12, the fitness function again detected an optimal path which avoided the larger Moderate terrain area.

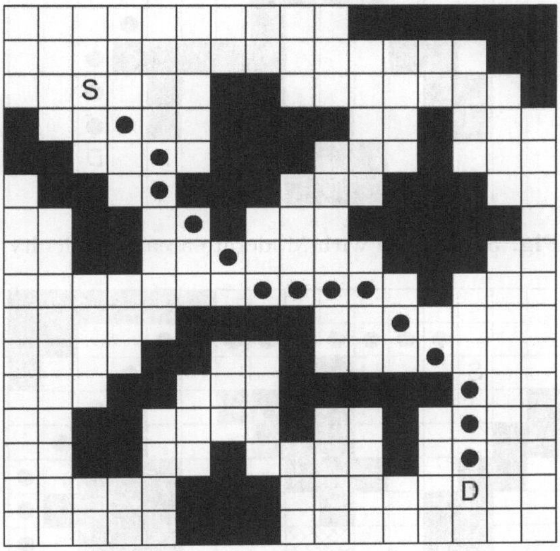

Fig. 3.9. Path generation with no terrain problems

A quadtree representation of the grid was also investigated for the same environment. The quadtree representation and resulting path are shown in Fig. 3.13. Tests demonstrate that an optimal path is also generated when using a quadtree environment. The only adjustment to the method is inclusion of a heuristic to decide to which cell to move in the event several exist when moving from a large aggregate cell to one of several adjacent smaller ones as shown in Fig. 3.14.

The system was also tested to determine its ability to respond in real-time. Since it is impractical to test every possible configuration of the environment grid, starting location, and destination, anecdotal evidence must be used. When tested with a environment grid size of 1024 by 1024, the robot was able to respond within 3 seconds for number of obstacle configurations. In a one square kilometer environment, this corresponds to a cell resolution of less than 1.0 meters – less than twice the diameters of the Pioneer 2-DX robot. This supports the efficacy of the real-time operation assertion for this approach.

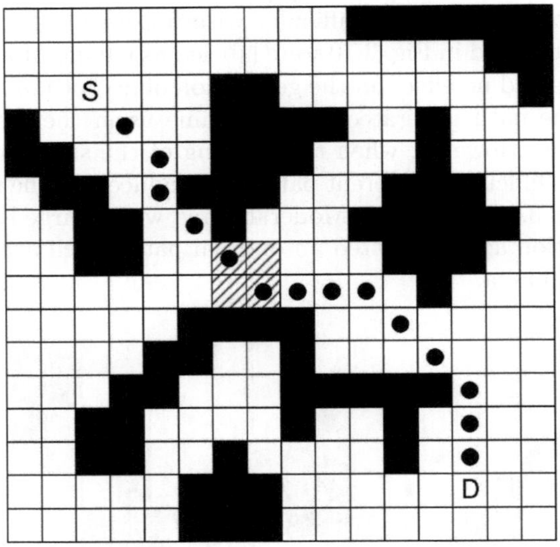

Fig. 3.10. Path with Moderate area of difficulty

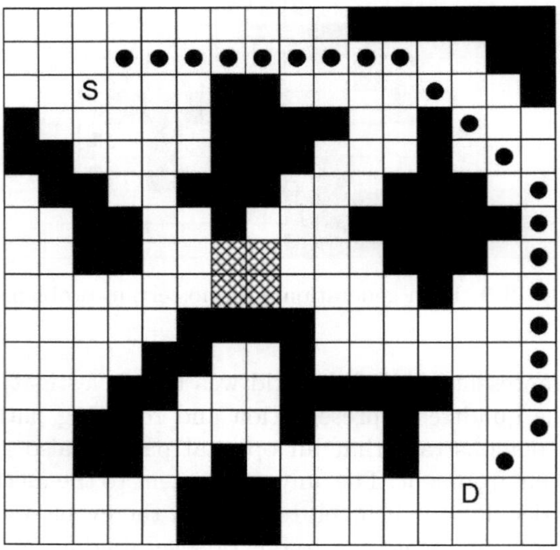

Fig. 3.11. Path with Difficult terrain area

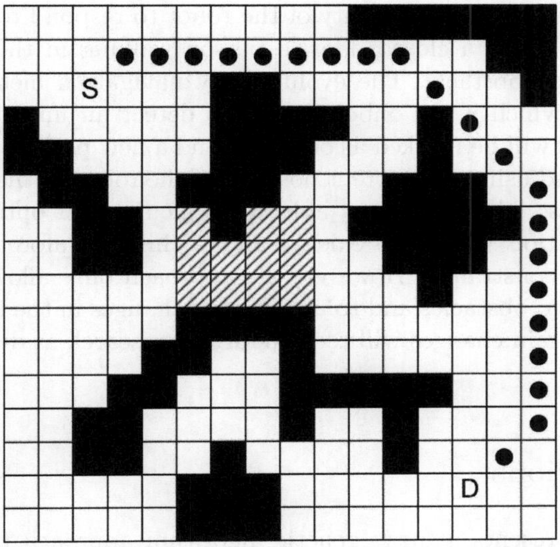

Fig. 3.12. Path with large area of Moderate difficulty

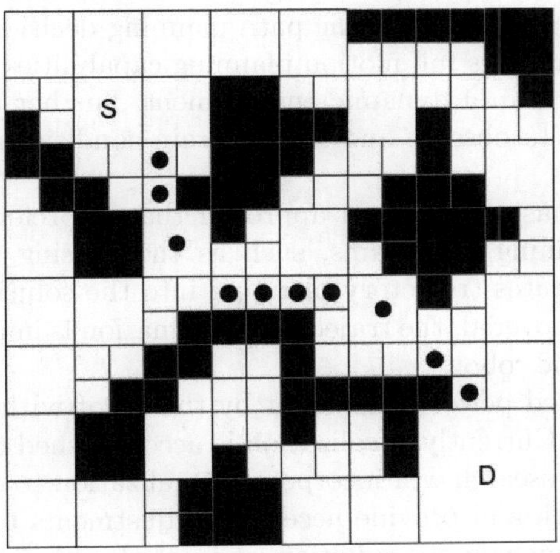

Fig. 3.13. Path using quadtree

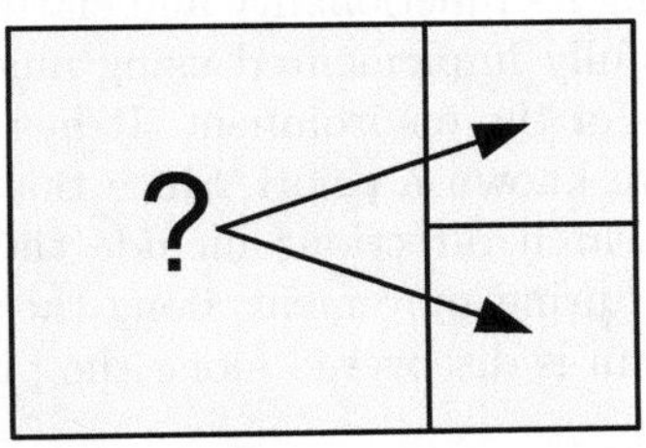

Fig. 3.14. Movement decision for quadtrees

Another concern was the ability of the robot to respond to changes in the environment, such as a closed door, a person walking in the area, or some other unexpected obstacle. The evolutionary navigation method provides a mechanism by which if the onboard sensors detect an unexpected obstacle, that cell or cells will be marked as occupied and a new path to the destination will be calculated using the current position of the robot as the starting point. In each test case, the robot was able to recalculate an optimal path from its current position when one existed and within the aforementioned real-time operation constraints. However, the approach only allows the robot to respond to sensed obstacles and to unexpected changes in the terrain changes. Response to terrain changes will require further research as discussed in Sect. 3.5.

3.5 Conclusions

This research presents a fuzzy genetic algorithm approach to motion planning for an autonomous mobile robot that performs in real-time without the limitations of monotone paths. Varying terrain conditions are represented as fuzzy values and are included in the path planning decision. The encoding of the chromosome provides full motion planning capabilities and the method is capable of operation in a dynamic environment. Further research directions include the ability to observe and learn terrain conditions during movement along the path.

This research has provided an approach that is preferable to many traditional path planning algorithms, such as those using search algorithms, because it incorporates trajectory planning into the solution. Thus, once an optimal path is discovered, the trajectory information is immediately available for movement of the robot.

We have assumed perfect movement by the robot without accounting for drift and slippage. Currently, localization is accomplished through dead reckoning. Additional research will incorporate localization to ensure the robot is on the planned path and provide necessary adjustments to the motion plan. Localization may include the addition of inertial and visual data. This research has presented the algorithm using a very simplistic 16 x 16 grid for purposes of demonstrating its functionality and clarity of the images. The approach has been successfully implemented using much larger grids and with quadtree representations of the environment. It has also been assumed that the terrain conditions are known a priori. Since this is not realistic in many applications, further research directions include the ability to observe and learn terrain conditions during movement along the path and to then adapt when more difficult terrain is discovered along the planned path.

References

1. Hwang Y K, Ahuja, N (1992) Gross motion planning – a survey. ACM Computing Surveys 24: 219–291
2. Latombe J C (1991) Robot motion planning. Kluwer, Boston Dordrecht London
3. Akbarzadeh M-R, Kumbla K, Tunsel E, Jamshidi M (2000) Soft computing for autonomous robotic systems. Computers and Electrical Egineering 26:5–32
4. Stafylopatis A, Bleka K (1998) Autonomous vehicle navigation using evolutionary reinforcement learning. European Journal of Operational Research 108:306–318
5. Zhang B-T, Kim S-H (1997) An evolutionary method for active learning of mobile robot path planning. In: Proceedings of the 1997 IEEE International Symposium on Computational Intelligence in Robotics and Automation. IEEE Press, New York
6. Fogel D B (2000) Evolutionary computation: toward a new philosophy of machine intelligence. IEEE Press, New York
7. Nearchou A C (1999) Adaptive navigation of autonomous vehicles using evolutionary algorithms. Artificial Intelligence in Engineering 13:159–173
8. Pratihar D K, Deb K, Ghosh A (1999) A genetic-fuzzy approach for mobile robot navigation among moving obstacles. International Journal of Approximate Reasoning 20: 145–172
9. Sugihara K, Smith J (1997) Genetic algorithms for adaptive motion planning of an autonomous mobile robot. In: Proceedings of the 1997 IEEE international symposium on computational intelligence in robotics and automation. IEEE Press, New York
10. Cordon O, Gomide F, Herrera F, Hoffmann F, Magdalena L (2004) Ten years of genetic fuzzy systems: current framework and new trends. Fuzzy Sets and Systems 141:5–31
11. Shibata T, Fukuda T (1993) Intelligent motion planning by genetic algorithm with fuzzy critic. In: Proceedings of the 8th IEEE symposium on intelligent control. IEEE Press, New York
12. Davidor Y (1991) Genetic algorithms and robotics: a heuristic strategy for optimization. World Scientific, London Singapore Hackensack
13. Nearchou A C (1998) Path planning of a mobile robot using genetic heuristics. Robotica 16:575–588
14. Iagnemma K, Dubowsky S (2004) Mobile robots in rough terrain. Springer, New York
15. Bonnafous D, Lacroix S, Simeon T (2001) Motion generation for a rover on rough tearrains. In: Proceedings of the 2001 IEEE international conference on intelligent robotics and systems. IEEE Press, New York
16. Hait A, Simeon T (1996) Motion planning on rough terrain for an articulated vehicle in presence of uncertainties. In: Proceedings of the 1996 IEEE/RSJ international conference on intelligent robotics and systems. IEEE Press, New York
17. Iagnemma K, Kang S, Brooks C, Dubowsky S (2003) Multi-sensor terrain estimation for planetary rovers. In: Proceedings of the 8th international symposium on artificial intelligence, robotics, and automation in space. IEEE Press, New York
18. Kelly A, Stentz A (1998) Rough terrain autonomous mobility – part 2: an active vision predictive control approach. Journal of Autonomous Robots 5:163–198

19. Pai D, Reissel L M (1998) Multiresolution rough terrain motion planning. IEEE Transactions on Robotics and Automation 14:19–33
20. Seraji H, Howard A (2002) Behavior based robot navigation on challenging terrain: a fuzzy logic approach. IEEE Transactions on Robotics and Automation 18:308–321
21. Shiller A, Chen J (1990) Optimal motion planning of autonomous vehicles in 3-dimensional terrains. In: Proceedings of the 1990 IEEE international conference on robotics and automation. IEEE Press, New York
22. Spero D, Jarvis R (2002) Path planning for a mobile robot in a rough terrain environment. In: Proceedings of the 3rd international workshop on robot motion and control. IEEE Press, New York
23. Davidson A, Kita N (2001) 3D simulation and map-building using active vision for a robot moving on undulating terrain. In: Proceedings of the 2001 IEEE computer society conference on computer vision and pattern recognition. IEEE Press, New York
24. Madjidi H, Negahdaripour S, Bandari E (2003) Vision-based positioning and terrain mapping by global alignment for UAVs. In: Proceedings of the IEEE conference on advanced video and signal based surveillance. IEEE Press, New York
25. Laubach S, Burdick J (1999) An autonomous sensor-based planner for microrovers. In: Proceedings of the 1999 IEEE international conference on robotics and automation. IEEE Press, New York
26. Lee T, Wu C (2003) Fuzzy motion planning of mobile robots in unknown environments. Journal of Intelligent and Robotic Systems 37:177–191
27. Jung I-K, Lacroix S (2003) High resolution terrain mapping using low altitude aerial stereo imagery. In: Proceedings of the 9th IEEE international conference on computer vision. IEEE Press, New York
28. Elfes A (1989) Using occupancy grids for mobile robot perception and navigation. IEEE Computer 22:46–57
29. Yahja, Stentz A, Singh S, Brumitt B (1998) Franed-quadtree path planning for mobile robots operating in sparse environments. In: Proceedings of the 1998 IEEE international conference on robotics and automation. IEEE Press, New York
30. Goldberg D E (1989) Genetic algorithms in search, optimization and machine learning. Addison-Wesley, Boston
31. Holland J H (1975) Adaption in Natural and Artificial Systems. University of Michigan Press, Ann Arbor

4

Aggregate Selection in Evolutionary Robotics

Andrew L. Nelson[1] and Edward Grant[2]

[1] Androtics LLC, PO Box 44065, Tucson, AZ 85733-4065 USA
alnelson@ieee.org
[2] Center for Robotics and Intelligent Machines, North Carolina State University, Raleigh, NC 27695 USA egrant@ncsu.edu

Can the processes of natural evolution be mimicked to create robots or autonomous agents? This question embodies the most fundamental goals of evolutionary robotics (ER). ER is a field of research that explores the use of artificial evolution and evolutionary computing for learning of control in autonomous robots, and in autonomous agents in general.

In a typical ER experiment, robots, or more precisely their control systems, are evolved to perform a given task in which they must interact dynamically with their environment. Controllers compete in the environment and are selected and propagated based on their ability (or fitness) to perform the desired task. A key component of this process is the manner in which the fitness of the evolving controllers is measured.

In ER, fitness is measured by a fitness function or objective function. This function applies some given criteria to determine which robots or agents are better at performing the task for which they are being evolved. Fitness functions can introduce varying levels of a priori knowledge into evolving populations. Some types of fitness functions encode the important features of a known solution to a given task. Populations of controllers evolved using such functions then reproduce these features and essentially evolve control systems that duplicate an a priori known algorithm. In contrast to this, evolution can also be performed using a fitness function that incorporates no knowledge of how the particular task at hand is to be achieved. In these cases all selection is based only on whether robots/agents succeed or fail to complete the task. Such fitness functions are referred to as *aggregate* because they combine the benefit or deficit of all actions a given agent performs into a single success/failure term.

Fitness functions that select for specific solutions do not allow for fundamentally novel control learning. At best, these fitness functions perform some degree of optimization, and provide a method for transferring known control heuristics to robots. At some level, selection must be based on a degree of

A. L. Nelson and E. Grant: *Aggregate Selection in Evolutionary Robotics*, Studies in Computational Intelligence (SCI) **50**, 63–88 (2007)
www.springerlink.com

overall task completion independent of particular behaviors or task solution features if true learning rather than simple optimization or transference is to be achieved.

Aggregate fitness functions measure overall task completion. However, they can suffer from an inability to produce non-random selection in nascent unevolved populations. If the task is too difficult, it is likely that none of the randomly initialized controllers will be able to make any meaningful progress toward completing the overall task.

This chapter investigates how aggregate fitness functions have been and continue to be used in ER, what levels of success they have generated relative to other fitness measurement methods, and how problems with them might be overcome.

4.1 Introduction

A distinction can be made between what a robot does and how it does it. That is, there is a difference between the task that a robot is to perform, and the manner in which it performs or solves the task. For example, consider a robot that is to be designed to move toward a light source (phototaxis). The robots task is phototaxis, but there are many ways in which this task could be performed. For instance, the robot might detect the light source, turn toward it, and then move forward until it collides with the source. Another solution might be for the robot to just wander around in its environment until it detected a threshold magnitude of light indicating it was near the source, at which point it would stop.

In general there are many solutions (of varying quality) to any given task. Determining the relative qualities of different solutions is essential for control learning. In artificial evolution-based forms of learning, fitness functions make this determination in an automatic or algorithmic way. The distinction between task and task solution defines two broad classes of fitness functions, namely *behavioral* fitness functions and *aggregate* fitness functions, and these will be a central focus of discussion in the following sections of this chapter.

Most autonomous robot systems are currently programmed by hand to perform their intended tasks. Learning to perform non-trivial tasks remains a largely unsolved problem in autonomous robotics. Evolutionary robotics approaches the problem of autonomous control learning through population-based artificial evolution. ER methods bear a great deal of similarity to other approaches to controller learning in autonomous robots. In particular, most learning methods require an objective function, and although the discussion in this chapter is focused on experimental work that involved population-based learning, most of the issues related to fitness evaluation are directly relevant to any autonomous system that is intended to learn basic control or behavior in a dynamic environment. We should point out here that other applications of machine learning in robots that are not aimed at learning the primary dynamic control of an agent, such as object recognition, mapping, or trajectory tracking, are generally amenable to heuristic algorithmic methods, or example-based gradient descent learning methods. These methods are not directly applicable to systems intended to learn how to perform complex tasks autonomously in dynamic environments.

To date, ER research has almost exclusively focused its attention on a handful of benchmark robot behavioral tasks, including phototaxis, locomotion and object avoidance, foraging, and goal homing. A large portion of this research has used fitness functions that selected for particular task solution elements that were known a priori to allow the task to be accomplished. Because of the simplicity of the tasks investigated, many of the fundamental problems associated with using fitness functions that contain a priori task solution information have not become apparent. If there are only a few near-optimal solutions to a particular problem, it is less obvious that a particular fitness function might have forced the evolution of some of the features of a resulting solution. The focus on a few simple tasks has also downplayed problems associated with the application of aggregate selection, to a degree. If a task is simple enough, an aggregate fitness function is likely to be able to drive evolution effectively starting from a randomly initialized population with no special treatment.

An important unanswered question within the field of ER is whether the methods used up to this point can be generalized to produce more sophisticated truly non-trivial autonomous robot control systems. Successful evolution of intelligent autonomous robot controllers is ultimately dependent on obtaining suitable fitness functions that are capable of selecting for task competence without specifying the low-level implementation details of those behaviors.

The remainder of this chapter is organized as follows: the rest of this section provides a general summary of ER methodology and related terms, and lays the foundation for discussion in later sections. In Sects. 4.2 and 4.3 we very briefly review the field of ER research in general and then provide a more in-depth review of ER work in which aggregate or nearly aggregate selection mechanisms were used. In Sect. 4.4 we discuss methods for overcoming difficulties associated with using aggregate fitness functions to evolve populations of mobile robot controllers for specific tasks. In Sect. 4.5 we provide an overview of our own work, which investigated methods for overcoming difficulties associated with aggregate selection in evolutionary robotics. In Sect. 4.6 we conclude the article with a discussion of the long-term prospects of using artificial evolution to generate complex autonomous mobile robot controllers.

4.1.1 Evolutionary Robotics Process Overview

In a typical ER experiment, artificial evolution and genetic algorithms are used to create robot or agent controllers able to perform some sort of autonomous task or behavior. The artificial evolution process starts with a randomly initialized population of robot or agent controllers. Over a series of generations or trial evaluation periods robots compete against one another to perform a given task. Robots that perform the task better are selected and their control systems are duplicated and altered using operations inspired by natural mutation and recombination. The altered controllers (the *offspring*) then replace the controllers of the poorly performing robots in the population. This process is iterated many times until a suitably fit controller arises.

It is possible to seed initial populations with controllers that are not randomly configured. Such populations might come from a previous evolutionary process (incremental evolution [1]), from a hand-directed training method (clicker or breeder training [2]), or could have been initially configured by some other means. However, such previously configured seed populations contain biases, and these biases can have an effect on the course of later evolution.

4.1.2 Bias

Before entering into a description of types of fitness functions used in ER, it makes sense to say a word or two about bias and sources of bias in evolving systems.

In the most general sense, bias refers to a tendency of a given dynamic system to develop toward a particular state. Bias in this context can be thought of as a form of pressure acting to change or evolve a system toward a particular quality or form. We use the terms *primary* and *secondary bias*.

Primary bias (or *representation bias*) describes any bias introduced due to a systems fundamental representation. Tendencies or abilities of a system due to the fundamental rules that describe the system (as in a simulation) are forms of representation bias.

Secondary biases are those that are imposed on a given system from outside the system in a way that is not necessarily consistent with the underlying representation of the system. Selection mechanisms in the form of fitness functions in artificial evolutionary systems represent secondary biases. As a side note we point out that selection mechanisms in natural evolution as observed in life on Earth do in fact stem at some level from primary representation bias, so in some sense there is no secondary bias in natural evolution. This is not the case in artificial evolutionary systems that use explicit selection mechanisms. Fitness functions in ER are imposed at a very high level and cannot be reduced to be consistent with fundamental physical law.

4.1.3 Fitness Functions

The fitness function is at the heart of an evolutionary computing application. It is responsible for determining which solutions (controllers in the case of ER) within a population are better at solving the particular task at hand. For optimization or classification applications, data sets and error minimization fitness functions can be applied and have proved to be very powerful. But these data sets and error minimization functions are not generally applicable to ER or intelligent control learning because the information required to formulate them is equivalent to the information the system is intended to learn. If one had in hand a functional specification of a given intelligent control algorithm at the sensor/actuator level, it would make little sense to use this to train an agent. A control program could be written directly from the specification. For all but the most trivial intelligent autonomous control problems, an appropriate mapping between sensor inputs and actuator outputs is not known. Often only a description of the robots final state in its environment is definable, hence this must be used to train the controllers. Standard methods of state space reinforcement learning (RL) such as Q-learning are also not applicable to non-trivial autonomous intelligent control problems because adequate discrete state spaces cannot be formulated. There has been some work developing methods of converting continuous state spaces into discrete state spaces, and to applying RL to continuous high dimensional systems, but again, the information necessary to formulate these representations is in general equivalent to the information needed to specify the desired control algorithm.

In current research aimed at evolving populations of autonomous robot controllers capable of performing complex tasks, the fitness function is almost always the limiting factor in achievable controller quality. This limit is usually manifested by a plateau in fitness in later generations, and indicates that the fitness selection function is no longer able to detect fitness differences between individuals in the evolving population.

As mentioned above, for purposes of discussion, will define two fundamental classes of fitness functions used in ER. These are *behavioral* and *aggregate*. We will also define a class of fitness functions that represents a combination of these two basic ones, taking elements from each. These will be called *tailored* fitness functions and are representative of a general tendency of researchers in this area to create hand-formulated fitness functions that select for many features of control that they might believe will result in successful evolution of controllers capable of performing a desired task.

Behavioral fitness functions are task-specific hand-formulated functions that measure various aspects of what a robot is doing locally and how it is doing it. These types of functions generally include several sub-functions or terms that are combined into a weighted sum or product. Behavioral fitness functions measure simple action-response behaviors, low-level sensor-actuator mappings, or low-level actions the robot might perform. For example, if one wished to evolve robots to move about an environment and avoid obstacles, one might formulate a behavioral fitness function that includes a term in the fitness selection function that is maximized if a robot turns when its forward sensors are stimulated at close range. In this example robot controllers will evolve to produce particular actuator outputs in response to particular sensor inputs. Selection occurs for a behavior that the designer believes will produce the effect of obstacle avoidance, but the robots are not evolving to avoid objects per se, they are learning to turn when their forward sensors are stimulated. This is more specific than just selecting for robots that do not collide with objects.

It is evident that if one does not have a basic understanding of how to perform a particular task, a behavioral fitness function cannot be formulated. But because many, if not all, tasks studied in ER to date are very simple, researchers have been able to formulate behavioral fitness functions using their own expertise and intuitions. For this reason behavioral fitness functions have been used extensively up to this point in ER. Examples of the use of behavioral fitness functions can be found in [3][4][5].

Some behavioral fitness functions are selective for a desired control feature, rather than a precise sensor-to-actuator mapping. For example, if one wished to evolve a robot controller that spent most of its time moving, one might include a term in the fitness function that is maximized when forward motion commands result in continued forward motion of the robot over time (if the front of a robot were in contact with an immobile object, it would not move forward regardless of its current actuator commands). This example term is not selective for an exact sensor-to-actuator mapping. There are many other formulations that could also produce the desired control feature. Hence, this type of term does not require quite as much a priori knowledge of the exact details of the control law to be learned.

For the evolution of non-trivial behaviors, selection using behavioral fitness functions results mainly in the optimization of human-designed controller strategies, as opposed to the evolution or learning of novel intelligent behavior. Because the tasks studied so far in the field have been relatively simple, and in many ways aimed at general proof of concept research, the reliance on behavioral fitness functions has not been such an important issue. However, it is safe to say that the groundwork in ER has been laid. It is possible to train robot control systems to perform behavioral tasks that require them to operate autonomously in dynamic environments.

Aggregate fitness functions select only for high-level success or failure to complete a task. Selection is made without regard to how the task was actually completed. This type of selection reduces injection of human bias into the evolving system by aggregating the evaluation of benefit (or deficit) of all of the robots behaviors into a single success/failure term. Evolution performed with an aggregate fitness function is sometimes called *all-in-one* evaluation. Until recently, aggregate fitness selection was largely dismissed by the ER community. This is because initial populations of controllers can be expected to have no detectable level of overall competence to perform non-trivial tasks (i.e. they are *sub-minimally competent*). Pure aggregate selection produces no selective pressure in sub-minimally competent populations at the beginning of evolution and hence the process cannot get started (the *bootstrap problem* [5]).

Even so, aggregate fitness selection in one form or another appears to be necessary in order to generate complex controllers in the general case if one is to avoid injecting restrictive levels of human or designer bias into the resulting evolved controllers.

Examples of aggregate fitness selection are found in [7][8][9]. Although there are many examples of the use of pure behavioral fitness functions, and some further examples of purely aggregating fitness functions, much of the ER research actually uses some form of hybrid between behavioral and aggregate selection. We will refer to these hybrid objective functions as *tailored fitness functions.*

Tailored fitness functions contain behavior-measuring terms as well as aggregate terms that measure some degree or aspect of task completion that is divorced from any particular behavior. As an example, suppose a phototaxis behavior is to be evolved. A possible fitness function might contain one term that rewards the degree to which a robot turns toward the light source, and another term that rewards a robot that arrives at the light source by any means, regardless of the specific sensor-actuator behaviors used to perform the task.

Unlike true aggregate fitness functions, aggregate terms in tailored fitness functions may measure a degree of partial task completion in a way that injects some level of a priori information into the evolving controller. For example, in the phototaxis task, a tailored fitness function might contain a term that provides a scaled value depending on how close the robot came to the light source during testing. This may seem at first glance to be free of a priori task solution knowledge or bias, but it contains the information that being closer to the goal is inherently better. In an environment composed of many walls and corridors, linear distance might not be a good measure of fitness of a given robot controller. We use the term tailored to emphasize that these types of fitness functions are task-specific hand-formulated functions that contain various types of selection metrics, fitted by the designer to the given problem, and often contain solution information implicitly or explicitly. Examples of work using tailored fitness functions can be found in [10][11][12].

To conclude our discussion of fitness functions we mention two variant methods of evolution that appear in ER. These are incremental evolution and competitive evolution.

Incremental evolution begins the evolutionary process by selecting for a simple ability upon which a more complex overall behavior can eventually be built. Once the simple ability is evolved, the fitness function is altered or augmented to select for a more complex behavior. This sequence of evolution followed by fitness function augmentation continues until eventually the desired final behavior is achieved. The overall process can be considered one of explicit training for simple sub-behaviors followed by training for successively more complex behaviors. In many ways this can be thought of as a serialization of a complex behavioral or tailored fitness function. The initial fitness functions in incremental evolution can be tailored or behavioral, but the final applied fitness function might be purely aggregate.

Competitive evolution (competitive selection) utilizes direct competition between members of an evolving population. Controllers in almost all ER research compete in the sense that their calculated fitness levels are compared during selection and propagation. However, in competitive evolution robot controllers compete against one another within the same environment so that the behavior of one robot directly influences the behavior, and therefore fitness evaluation, of another.

A variant upon competitive evolution is co-competitive evolution in which two separate populations (performing distinct tasks) compete against each other within the same environment. Examples of co-competitive evolution involving populations of predator and prey robots exist in the literature [13][14][15]. Two co-evolving populations, if initialized simultaneously, stand a good chance of promoting the evolution of more complex behaviors in one another. As one population evolves greater skills, the other responds by evolving reciprocally more competent behaviors. The research presented in [13][14][15] shows this effect to a degree, but results from other areas of evolutionary computing suggest that given the correct evolutionary conditions, pure aggregate selection combined with intra-population competition can result in the evolution of very competent systems [16][17].

4.2 Evolutionary Robotics So Far

The field of ER has been reviewed in several publications [18][19][20]. Much of the research focuses on evolving controllers for simple tasks such as phototaxis [21][22], object avoidance [23][24], simple forms of navigation [25][26], or low-level actuator control for locomotion [5][27].

Although the evolutionary algorithms vary to a degree from work to work, most of them fall within a general class of stochastic hill-climbing learning algorithms. Unless otherwise stated, the research discussed below use some form of evolutionary algorithm roughly equivalent to that which was outlined in the introduction to this chapter. It is true that some algorithms may show a two-fold (or even ten-fold) increase in training efficiency over others, but so long as the search space and controller representation space are not over-constrained it is the fitness function that finally determines the achievable performance.

We consider mainly evolutionary robotics work that has been verified in real robots. Physical verification in real robots forces researchers to use simulations that are analogous to the physical world. Some subtleties of control are contained within the robot-world interface and are easily overlooked [28][29]. In particular, adequate simulators must maintain a suitable representation of the sensory-motor-world feedback loop in which robots alter their relationship to the world by moving, and thus alter their own sensory view of the world. Robotics work involving only simulation, without physical verification, should not be considered fully validated. Much of the pure-simulation work falls into the category of artificial life (AL), and many of these simulation environments include unrealistic representations or rely on sensors that report unobtainable data or conceptual data. That said, learning in simulation with transfer to real robots has been repeatedly demonstrated to be viable over the last decade [8][30][31][32]. Much of this work has involved new physics- and sensor-based simulators. The verification of evolved controllers in real robots allows a clear distinction to be made between the large amount of work done in the field of artificial life, and the similar, but physically grounded work pursued in evolutionary robotics.

Early research efforts that might be considered precursors to evolutionary robotics done in the late 1980s consisted of learning simple navigation abilities in purely simulated agents [28][33]. In the first decade of evolutionary robotics work (1990–2000), the field moved from an almost non-existent state to become a field of research in which numerous projects produced real robots relying entirely on evolved controllers for interaction with their environments.

Locomotion in combination with obstacle avoidance in legged robots has been reported in several studies [1][5][23][7]. Filliat et al. [1] evolved locomotion and object avoidance controllers for a hexapod robot using neural networks composed of threshold neurons. Controllers were evolved in simulation and transferred to real robots for testing. Jakobi et al. [5] described the use of minimal simulation to evolve controllers for an eight-legged robot with sixteen leg actuators. Kodjabachian et al. [23] describe the incremental evolution of walking, object avoidance and chemotaxis in a simulated six-legged insectoid robot. Hornby et al. [7] describe the evolution of ball chasing using an 18-DOF quadruped robot.

Peg pushing behaviors were evolved in [6][34]. This task required robots to push small cylinders toward a light source. In [35] Lee et al. investigated a similar box-pushing behavior using Genetic Programming (GP).

Several examples of competition in the form of co-evolution of competing species have been reported in the literature. Cliff and Miller investigated the co-evolution of competing populations of predator and prey robots [15][36]. Similar works have been reported in [13][14][19].

Evolution of controllers using competition within a single population (intra-population competition) is investigated in [32].

The most complex tasks addressed in the literature involve some form of sequential action. Nolfi [37] reports on the evolution of a garbage collection behavior in which a robot must pick up pegs in an arena and deposit them outside the arena. Ziemke [38] studied the evolution of robot controllers for a task in which a robot must collide with objects (collect them) in one zone and avoid them in another. In [39] Floreano et al. report on the evolution of a behavior in which robots move to a light and then back to a home zone. Another example of evolving controllers for a relatively complex task is reported in Tuci et al. [40]. Robot controllers evolved to produce lifetime learning in order to predict the location of a goal object based on the position of a light source.

Flocking behaviors have also been investigated. Ashiru describes the evolution of a simple robot flocking behavior in [41]. A robot coordination task in which two robots evolve to move while maintaining mutual proximity is reported by Quinn in [42]. Baldassarre et al. [43] evolved homogeneous controllers for a task in which four robots must move together in a small group toward a light or sound source. In [44] aggregation of small robots into a larger structure is investigated and makes use of a relatively complex hand-formulated fitness function.

In the early 2000s evolution of body and mind (morphology and controller) was achieved using the innovation of elemental modular components that were both amenable to simulation, and relatively easy to fabricate [8][30]. It must be noted that these works have not produced significant advances in controller complexity per se, but rather they have shown that evolution in simulated environments can indeed be used to produce physically viable robot minds and bodies.

It may still be the case that certain environments are beyond the ability of modern methods to simulate for the purpose of evolutionary learning, but this cannot now be considered to be the stumbling block that it once was. Numerous experiments and systems have shown that the intuitively compelling arguments supporting embodiment as a requirement for low-level learning entertained by researchers in the late 1980s and early 1990s [29] are in fact not correct [8][30][31][32]. The universe is by no means its own best simulation[28].

Recent years have also seen a significant methodological change related to fitness function usage in the field. Aggregate fitness functions have been used to reproduce many of the results first obtained using more complicated hand-formulated fitness functions. Some earlier works, especially in the area of incremental evolution, made the claim that the more complex forms of fitness evaluation were necessary to achieve successful evolution of the behaviors studied. This however has been shown not to be the case.

Although developing an experimental research platform capable of supporting the evolutionary training of autonomous robots remains a non-trivial task, many of the initial concerns and criticisms regarding embodiment and transference from simulated to real robots have been addressed. There are sufficient examples of evolutionary robotics research platforms that have successfully demonstrated the production of working controllers in real robots [13][21][22][24]. Also, there have been numerous examples of successful evolution of controllers in simulation with transfer to real robots [3][7][8][37][45][46][47].

One of the major achievements of the field of ER as a whole is that it has demonstrated that sophisticated evolvable robot control structures (such as neural networks) can be trained to produce functional behaviors in real (embodied) autonomous robots. What has not been shown is that ER methods can be extended to generate robot controllers capable of complex autonomous behaviors. In particular, no ER work has yet shown that it is possible to evolve complex controllers in the general case or for generalized tasks.

Concerns related to fitness evaluation and fitness selection remain largely unresolved. The majority of ER research presented in the literature employs some form of hand-formulated, task-specific fitness selection function that more or less defines how to achieve the intended task or behavior. The most complex evolved behaviors to date consist of no more than three or four coordinated fundamental sub-behaviors [37][31][32][39]. In [37], the fitness selection method used was relatively selective for an a priori known or pre-defined solution. In [31][32][39] the fitness functions used for selection contained relatively little a priori knowledge, and allowed evolution to proceed in a relatively unbiased manner. This is an interesting contrast to much of the work aimed at evolving simple homing or object avoidance behaviors, which in some cases used complex fitness functions that heavily biased the evolved controllers toward an a priori known solution.

4.3 Evolutionary Robotics and Aggregate Fitness

In this section we focus on evolutionary robotics research that has used aggregate fitness functions.

Several robot actuator control tasks have been investigated using aggregate or near-aggregate fitness functions. These include gait evolution in legged robots [9][48][49][50], and flying lift generation in a flying robot [27]. Simple actuator coordination tasks do not fall under the heading of environmentally situated intelligent autonomous robot control and do not usually produce complex reactions to environmental sensor stimuli. However, they do involve the application of evolutionary computing methods to evolve novel control, and are included in this section along with the other evolved autonomous controller research.

In [7] the evolution of a ball-pushing behavior using an 18-DOF quadruped robot (Sony AIBO) is described. The fitness function used can be considered to be aggregate. The function measures the degree of success of moving the ball simply by measuring the total distance that the ball was moved over the course of an evaluation trial.

In [48] the authors use embodied evolution to develop gaits for a hexapod robot. An aggregate fitness function was used that measured the distance traveled by the robot while walking on a treadmill.

Both [8] and [30] described separate examples of systems in which whole robots (bodied and controllers) were co-evolved in simulation and then constructed in the real world using modular actuators and structural units. In both cases robots were evolved for locomotion abilities and fitness was calculated simply as the distance d traveled. This was a purely aggregate fitness function and contained no other features of potential control solutions or of possible robot morphologies.

Reference [9] reported on the embodied evolution of a locomotion behavior in a robot relying on tactile sensors for object detection. The fitness function simply measured the arc-length distance traveled by the robot (as reported by a swivel wheel and odometer attached to the robot) over a given evaluation period.

In [27] embodied evolution was used to develop lift-generating motions in a winged flying robot. A near aggregate fitness function was used that measured height obtained by the robot at each time step.

In [51] an indoor floating robotic blimp equipped with a camera and placed in a small room with bar-code-like markings on the walls was evolved to produce motion and wall avoidance. An essentially aggregate fitness function was used that averaged magnitude of velocity over each trial period.

Reference [52] described the evolution of morphology and control for modular robots constructed of Lego and servo units. Robots were evolved for locomotion abilities in simulation and then constructed in the lab with real hardware. A near-aggregate fitness function (the same as that used in [8] and [30]) was used that measured total net locomotion distance over the course of a trial period. Note that an initial settling period occurred before each fitness-measuring period began. This was done to avoid selecting for robots that moved merely by falling and this makes the fitness function technically tailored, to a small degree.

Reference [49] presents another example of embodied evolution of gaits in a physical robot. The robot was a pneumatic hexapod of minimalist design. The authors used the same aggregate fitness function as did [48]. Distance for the aggregate fitness function was determined using images taken from an overhead camera.

Gait learning using a slightly different aggregate fitness function is given in [50]. In this paper, a Sony AIBO quadruped robot was used, and again, the evolutionary process was embodied in the real robot. Here fitness was measured as average speed achieved by the robot.

4.4 Making Aggregate Selection Work

This section discusses methods for overcoming difficulties associated with using aggregate fitness functions to evolve populations of mobile robot controllers for specific tasks.

The foremost problem with aggregate selection is that it lacks selective power early in evolution. One way to address this is to use a behavioral or tailored fitness function to train robots to the point at which they have at least the possibility of achieving a given complex task at some poor but detectable level, and then to apply purely aggregate success/failure selection alone in the later part of evolution, thus relaxing biases introduced by the initial fitness function. The term *bootstrap mode* is used to refer to such an initial training function.

It is not clear what legacy in the evolving population might be left by biases introduced by an initial bootstrap mode fitness function. Once started down one path, innovation might be inhibited even if the initial biases are removed. Also, designers still need to have some idea what kinds of fundamental abilities are needed to allow the evolving controllers to have some chance at completing the task. Even with these problems, the use of an initial bootstrap mode in conjunction with later pure aggregate selection is a viable method for some non-trivial tasks and will likely be studied a great deal in the coming years.

Using intra-population competition in conjunction with aggregate selection may also improve the quality of evolved controllers, at least for tasks that are inherently competitive. Competitive fitness selection utilizes direct competition between members of an evolving population. Controllers in almost all ER research compete in the sense that their calculated fitness levels are compared during selection and propagation. However, in competitive evolution robot controllers compete against one another within the same environment so that the behavior of one robot directly influences the behavior, and therefore fitness evaluation, of another. For example, in a competitive goal-seeking task, one robot might keep another from performing its task by pushing it away from the goal. Here, the second robot might have received a higher fitness rating if it had not been obstructed by the first robot.

Intra-population competition presents a continually increasing task difficulty to an evolving population of controllers and may be able to generate controllers that have not been envisioned by human designers.

4.5 Aggregate Selection and Competition

In this section we provide an overview of our own ER research [32][46]. The specific aim of the experiments reviewed here was to extend aggregate selection using the methods discussed in the previous section.

We used aggregate selection with a minimal bootstrap mode in conjunction with direct intra-population competition. The bootstrap mode was triggered only when no controller in the current generation of the evolving population showed any detectable ability to complete the overall task.

As in the majority of ER research, we used neural networks to control our robots. Populations of neural network based controllers were evolved to play a robot version of the competitive team game *Capture the Flag*. In this game, there are two teams of mobile robots and two stationary goal objects. All robots on one team and one of the goals are of one color (red). The other team members and their goal are another color (green). In the game, robots of each team must try to approach the other teams goal object while protecting their own goal. The robot which first comes within a range of its opponents goal wins the game for its team. Winning the game here is the task the robots learn to solve. The game is played in maze worlds of varying configurations.

The competitive task, in the form of a game, allows for both simple and complex strategies to arise. Although the simplest solution to this game task that might succeed is to wander about the environment until the opponents goal is found, more complex strategies are more efficient. Such a simple strategy will usually fail against a more competent opponent. The competitive element of the evolutionary environment is needed to drive controllers toward better solutions, and this is where our research differs from most other research in the field. We employ direct competition during evaluation between robots. This means that the actions of one robot can alter the fitness evaluation of another directly. This also produces a changing fitness landscape over the course of evolution (the Red Queen Effect [36]).

The neural networks are essentially blank slates at the beginning of evolution. They are made up of randomly interconnected neurons with weighted connections initialized using values from a random distribution and they contain no information related to the task to be learned. In addition to learning any game-specific behaviors, the robots must also learn all aspects of locomotion and navigation. The robot controllers relied solely on processed video inputs for sensing their environment, and the best-performing neural network controllers contained on the order of 100 neurons and 5000 connections. Because of the very large number of sensor inputs, and the number of different types of objects the robots needed to learn to recognize (or at least respond to), the actual task learned by the controller is in some ways much more difficult than those studied in other related work.

The genetic representation of the neural controllers was a direct encoding of the network weights and connection topology. This consisted of a matrix of real-valued elements in which each element value represented a connection weight, and the location of each element represented the position of that connection in the overall network. The matrix also contained several additional columns of formatted fields that specified neuron types and time delays.

During evolution, only mutation was used. Network weights, connectivity and network topology were evolved. For the work discussed below, populations of 40 individual controllers were evolved. The evolutionary conditions and parameters are summarized in Table 4.1 below.

The physical robots used in this work were the EvBots [1][41]. The robots were fully autonomous and performed all vision processing and control computation on board. Figure 4.1 shows a photograph of two EvBots. Each robot has been fitted with a colored shell. The shells were used in the *Capture the Flag* game behavior and served to differentiate robots on different teams.

Testing of the robot controllers was performed in the reconfigurable CRIM robot maze at the Center for Robots and Intelligent Machines at North Carolina State University (Fig. 4.2). The controllers themselves were evolved in a simulation environment coupled to the real robots and maze environment at the sensor and actuator level [46].

Fitness for individual controllers was based on their performance in competition in tournaments of games. During each generation, a single tournament of games was played. A bimodal training fitness selection function was used. The fitness function has an initial bootstrap mode that accommodates subminimally competent seed populations and a second mode that selects for

Table 4.1. Environmental and experimental parameters used during evolution

Parameter	Setting
Population size	40
Sensor inputs	150
Initial network size	60 neurons
Chance of adding or removing a neuron (during mutation)	70%
Weight initialization range	[-1 1], uniform distribution
Weight mutation magnitude	[-1 1], uniform distribution
Weight mutation rate	25%
Initial feed forward connectivity	60%
Initial feedback connectivity	20%
Chance of adding or removing a connection (during mutation)	70%
Elitism level (per generation)	Single best from previous generation
Population replacement rate	50%
Generations (per evolution)	650

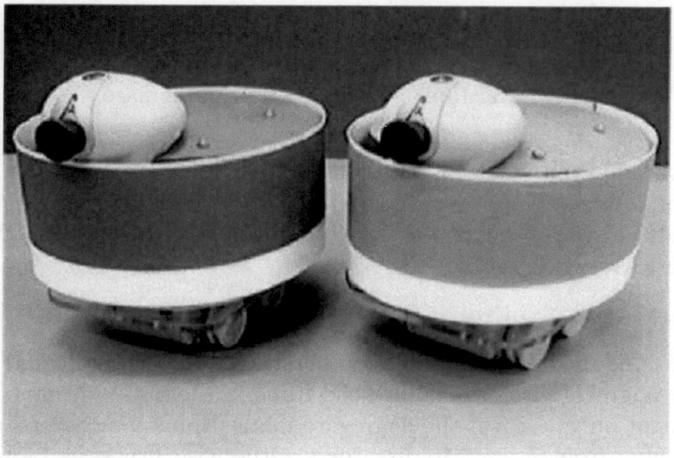

Fig. 4.1. EvBot robots fitted with colored shells

aggregate fitness based only on overall success or failure (winning or losing games). The triggering of modes was decoupled from any explicit aspect of the training environment or specific generation number and was related only to the current behavior of the population.

The fitness function was applied in a relatively competitive form in which controllers in the evolving population competed against one another to complete their taskto win the game. In a given generation, if any robot controller was able to complete the task, then all information from the bootstrap mode was discarded for all the robots regardless of their individual performances

Fig. 4.2. Robots in the CRIM testing maze

and selection for the whole population was conducted with purely win/lose information.

Formally, fitness $F(p)$ of an individual p in the population $\mathbf{P}$ at each generation was calculated by:

$$F(p) = F_{mode_1}(p) \oplus F_{mode_2}(p) \tag{4.1}$$

where F_{mode_1} is the initial minimal-competence bootstrap mode and F_{mode_2} is the purely aggregate success/failure-based mode. Here $\oplus$ indicates exclusive-or, dependent on F_{mode_2}: if the aggregate modes value is non-zero, it is used and any value from F_{mode_1} is discarded. Otherwise fitness is based on the output of F_{mode_1}. F_{mode_1} is formulated to return negative values and returns 0 when maximized or if F_{mode_2} is active. F_{mode_2} in contrast returns positive values based on number of game-wins, if any, and this allows for a simple mechanism to track which mode is active at any generation over the course of evolution.

The minimal competence bootstrap mode selects for the ability to travel a distance D through the competition maze environment. The general form of mode 1 is as follows:

$$F_{mode_1}(p) = F_{dist} - s - m \tag{4.2}$$

where F_{dist} calculates a penalty proportional to the difference between distance d traveled by the best robot on a team, and the minimal competence distance D, which is defined as half the length of the training environments greatest dimension. In (4.2), s and m are penalty constants applied when robots on a team become immobilized or stuck (by any means), or when controllers produce actuator output commands that exceed the range of the actuators (the wheel motors) respectively.

The rationale for this particular bootstrap mode is that if a robot can navigate at least partway through a given environment without becoming ensnared on walls, other robots, or other obstacles, then it has some chance of running across its quarry (i.e. the opponent goal).

The second mode of the fitness function F_{mode_2} is classified as aggregate because it produces fitness based only on success or failure of the controllers to complete the overall task (i.e. winning the game by finding and touching the opponents goal first). The formulation of the success/failure mode of the fitness function is determined by the competitive nature of the training algorithm. In each generation, a tournament of games involving all the individuals in the population was conducted. Each individual played two games against another member of the population (the opponent). Note that the opponent was selected at random from the previous generation of the population at the beginning of each tournament and all controllers competed against that one robot. The reason for this was to reduce the stochastic differences during evaluation introduced by the environment and random opponent selection. The possible outcomes of these games incurred different levels of fitness and are summarized in Table 4.2 below.

Table 4.2. Fitness points awarded by the aggregate success/failure mode F_{mode_2}, for pairs of reciprocal games during a generational tournament

Game Pair Outcomes	Fitness Points Awarded
win-win	3
win-draw	1
win-lose	0.5

Note that in cases where no win occurs during the entire tournament F_{mode_1} is used to determine negative fitness values.

Over the course of a typical controller evolution run, the fitness function started using purely bootstrap mode, progressed to using a mixture of bootstrap and aggregate, and then in the later parts of evolution relied exclusively on pure aggregate selection. Figure 4.3 shows fitness values and number of games won per generation over the course of an evolutionary run. The active mode of the fitness function used during each generation is indicated at the top of the figure. No controller in the population was able to win a game before the 60th generation of training. With a population size of 40 (a typical size for this research) this would represent about 4000 games and indicates that the initial and early forms of the population have virtually no ability to complete the overall task and that use of the aggregate fitness mode alone would indeed fail. Between the 60th and 160th generations the fitness function oscillates between the two modes, relying more heavily on the pure aggregate mode as evolution continues. After the 160th generation the bootstrap mode is not invoked again, and selection is based completely on aggregate selection. Note that the differing fitness values correspond to the fitness values listed in Table 4.2 when the aggregate mode is active. In addition, note that the incremental climb in fitness in the first 60 generations when fitness is dominated by the bootstrap mode is typical of ER trainings in which a behavioral fitness function is used. Further, as the aggregate fitness mode becomes dominant, absolute fitness becomes less of an indicator of population refinement. Since the high score is 3 points, any generation that has a controller that can win two games it plays will have this as the best fitness.

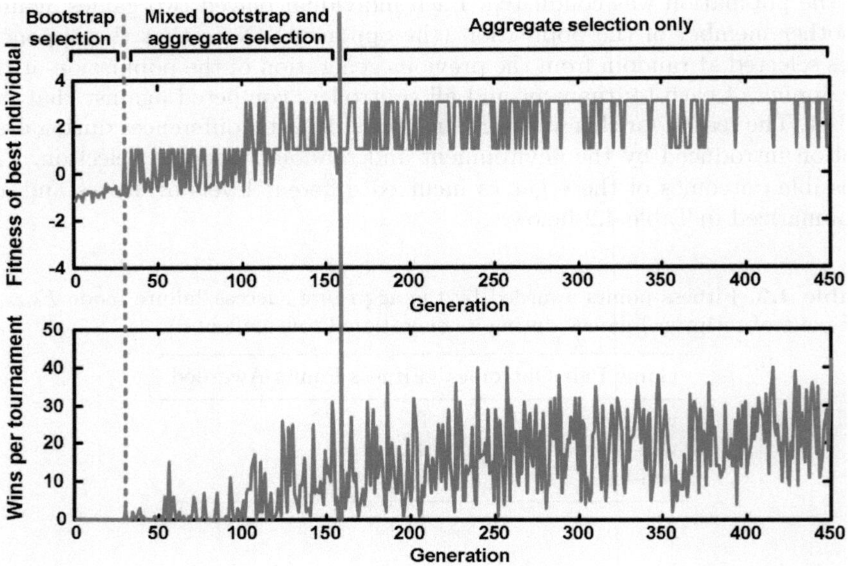

Fig. 4.3. Fitness of best individual and number of wins per generation are shown over the course of evolution for an evolving population. Modes of fitness function operation are also indicated

Over the course of evolution the robots learned how to navigate and locate goals. They could differentiate between the different types of objects in their environment as judged by their different responses to these. The robot controller also evolved some temporal behaviors, but relied mainly on reactive control.

The results shown in Fig. 4.3 summarize data collected over a particular evolutionary training session. A series of such evolutions were performed, and produced generally similar results, but due to very long simulation times (on the order of three weeks of computation time on a 3.2 GHz X86 class dual processor machine) no statistically significant data summarizing a putative general case were generated.

Example games played with evolved controllers are shown in Fig. 4.4. The first panel shows a screen shot of the simulation environment while the second panel shows a view of the physical maze environment and the real robots taken with an overhead video camera. In both of the example games the robots are shown in their final positions at the end of the games. The paths taken by the robots are indicated by lines superimposed on the images.

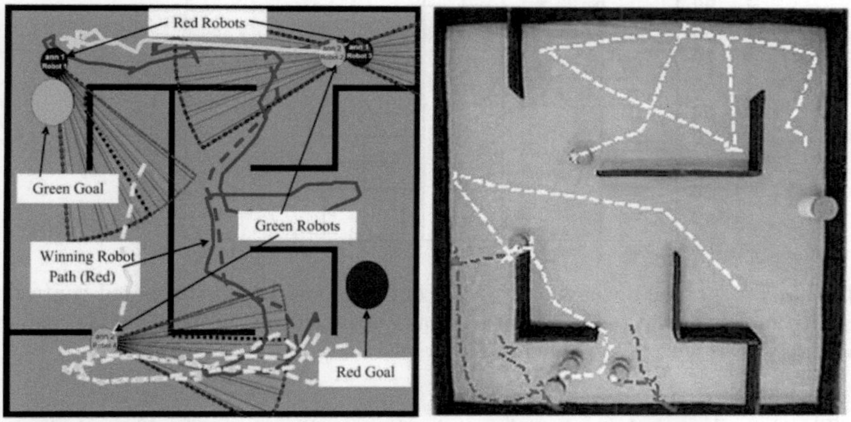

Fig. 4.4. Games played with robots using evolved controllers in simulated and real environments

The game sequences of Fig. 4.4 demonstrate that controllers have evolved some degree game-playing behavior. Because a relatively competitive fitness selection metric was used to drive the evolutionary process, the absolute quality of the robot controllers is not known. To address this, after the evolutionary training process was complete, the final evolved controllers were tested against a hand-coded knowledge-based controller of well-defined abilities.

The hand-coded controller was not used in any way during the training of the neural network-based controllers. As far as the evolved controllers were concerned, the rule-based controller was a novel controller not seen during training. In addition, the post-training games were played in a maze configuration not used during training in order to rule out the possibility that the controllers memorized a particular maze topology as part of their learned strategies.

In order to obtain a reliable result a series of 240 games between the rule-based controller and the best trained neural network controller was conducted in an environment similar to the ones shown in Fig. 4.4. Each game during the tournament was initialized with a new randomly generated set of starting positions for robots and goals. Figure 4.5 shows the results of this tournament. The best-evolved neural controller won 108 games, the knowledge-based controller won 103 games, and 29 games were played to a draw.

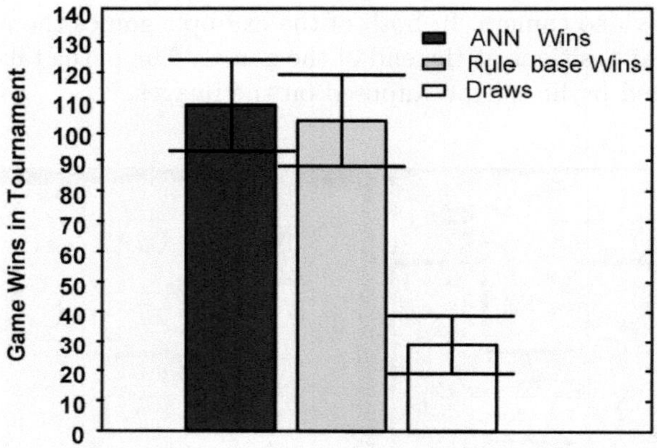

Fig. 4.5. Bar graphs displaying evolved controller and knowledge-based controller competition data collected during a tournament of 240 randomly initialized games. Data are shown with 95% confidence intervals

The data do not show that the evolved controllers were significantly better than the hand-coded controller. Even with the large number of games played, the total number of wins for each type of controller overlapped within 95% confidence intervals. The evolved controllers played competitively with the hand-written controllers, though, and this result is statistically significant.

In marked contrast to the knowledge-based controllers, the neural network-based controllers displayed complex trajectories that were extremely difficult to predict exactly. Although it may be possible to qualitatively analyze the evolved controller behaviors to a degree, such analysis is not at all necessary to the evolutionary process. Competition alone during evolution was responsible for driving the complexity of the controllers to a level at which they could compete the hand-coded controller. Human knowledge was required to formulate the minimal competence mode of the fitness function, but this only selected for minimal navigation abilities.

The work discussed above used a bootstrap selection mode early in evolution before transitioning to an aggregate mode. This raises the question of how these experiments compare to similar work that used only purely aggregate selection from the beginning. For comparison, we attempted to use purely aggregate selection without the initial bootstrap mode. For our particular competitive game, purely aggregate selection was not able to drive the evolution of fit controllers (data not shown). However, in [53], carefully structured environmental-incremental evolution, in which the environment used during training was made progressively more difficult, did produce competent controllers in a simulated experiment. The task in [53] required robots to find and pick up objects in an arena, and then deposit them outside the arena. The majority of the other evolutionary robotics work reported in the literature that used purely aggregate fitness functions evolved simple locomotion behaviors and is not directly comparable to more complex tasks.

In [16] networks for playing Checkers were evolved to the near-expert level (able to beat 95% of human players) using a purely aggregate selection scheme. This was a non-robotic task, and the game of Checkers lacks the dynamic sensor-motor-environment feedback loops characteristic of autonomous robotic systems, but the work demonstrated that it is possible to evolve very competent systems for difficult tasks.

4.6 Conclusion

We conclude the chapter with a discussion of the long-term prospects of using artificial evolution to generate highly complex autonomous mobile robot controllers. In spite of recent advances in ER, there remain fundamental unsolved problems in the field that prevent its application to many non-trivial real-world problems.

Although the field of evolutionary robotics has expanded and developed greatly in the past decade, in some sense no substantive improvement in overall controller sophistication has been achieved during this time. This is in spite of continued advances in computer speeds, availability of robot platforms, and efficiency of algorithms.

We hypothesize that the lack of improvement in the field of ER is fundamentally a result of an inability to formulate effective fitness functions for non-trivial tasks. For complex tasks, behavioral fitness functions must contain more and more features of a known solution. In effect, the fitness function becomes a recipe for evolving a solution containing mainly a priori known features and more or less defines all of the major features of the evolved solutions.

We believe that aggregate selection must be used in one form or another to achieve complex behaviors that are not known by researchers and developers a priori, and this is crucial for the application of ER to sophisticated robot learning in dynamic uncharacterized environments. This is not just a problem for ER, but affects any form of learning in uncharacterized environments that require some form of feedback to drive the learning process.

The work in [16] and also in [17] use aggregate selection to generate very competent systems for playing the games of Checkers and Go respectively.

However, the current robotics work has not reached these levels of proficiency. Evolution conducted with one or more carefully formulated bootstrap modes, followed by aggregate selection performed in a fidelity training environment with adequate computing power will very probably be able to generate controllers an order of magnitude more complex than the current standard fitness evaluation methods in ER. But this is in no way a general solution to autonomous control learning. Controllers created using most current ER methods might be capable of performing one or two distinct elemental tasks, and the most sophisticated controllers evolved perform no more than five or six coordinated elemental tasks. An order of magnitude improvement would not represent a significant improvement over current non-learning-based autonomous robot controllers, although it would allow ER methods to at least be competitive with other state-of-the-art autonomous robot control design methods.

Intra-population competition during fitness evaluation, in combination with aggregate or near-aggregate selection, can produce controllers of significant complexity, but many if not most tasks of interest cannot be reduced to a purely competitive form in which one robots behavior directly impinges upon the behavior (and thus the fitness) of another co-evolving robot. This direct intra-population competition within the robots testing environment is essential in order to drive evolution to produce sophisticated agents, at least if it is applied as in the experiments discussed in this chapter. So again, we do not yet have a tool that can provide generalized learning for most non-trivial problems.

A second and more esoteric problem with aggregate selection, as well as other selection mechanisms in artificial evolving systems, is that they impose a particularly high-level of task bias on the evolving agents in a way that cannot be related to natural evolutionary processes. This is not to say that every aspect of natural evolution must be duplicated in artificial evolution systems, but because the natural development of life on Earth appears to be our only example of truly complex agents arising from lifeless origins, it is unclear which aspects of natural evolution are essential, and which are merely artifacts of life as it developed on Earth.

Artificial evolution when applied to autonomous robots attempts to evolve abilities to perform specific tasks. In natural systems, there is no particular fundamental bias toward any particular functionality. This is an important and perhaps subtle difference between natural and artificial evolution. Natural evolution does not evolve creatures with specific abilities per se. Nature used (uses) exactly the same fundamental driving selection force to evolve, for example, trees as it did to evolve birds. The underlying bias toward particular functionality (i.e. photosynthesizing or flying) in natural systems is very low. Further, all forms of bias seen in nature are reducible to fundamental physical law (even if this reduction is beyond our abilities to elucidate). Because representation (i.e. the physical universe) and selection pressure in nature are consistent, fundamental bias is very low, yet complexity can be represented.

It is possible that in order for anything approaching life-like abilities to arise, all forms of selection bias must be kept at extremely low levels, and must be consistent with fundamental representation. This places into question whether it is possible to evolve agents for particular predefined tasks in general.

There are no artificial evolutionary systems in which selection and representation are fully consistent. Currently, all artificial life systems, including all evolutionary robotics systems, that have shown any signs of developing autonomous capabilities use some form of fitness evolution that is imposed on the evolving systems. Self-organizing systems such as cellular automata have been studied, and these perhaps do not suffer from secondary biases, but no such system has been shown to have the potential to generate intelligent interactive behaviors that are necessary for autonomous robotic systems that are intended to operate in dynamic environments.

Acknowledgment: The authors would like to thank Brenae Bailey for editorial assistance and insightful input related to this work.

References

1. Filliat D, Kodjabachian J, Meyer J A (1999) Incremental evolution of neural controllers for navigation in a 6 legged robot. In: Sugisaka, Tanaka (eds) Proc. Fourth International Symposium on Artificial Life and Robotics. Oita Univ. Press
2. Lund H H, Miglino O, Pagliarini L, Billard A, Ijspeert A (1998) Evolutionary robotics – a children's game. In: Evolutionary Computation Proceedings, 1998 IEEE World Congress on Computational Intelligence
3. Lund H H, Miglino O (1996) From simulated to real robots. In: Proceedings of IEEE International Conference on Evolutionary Computation
4. Banzhaf W, Nordin P, Olmer M (1997) Generating adaptive behavior using function regression within genetic programming and a real robot. In: Proceedings of the Second International Conference on Genetic Programming. San Francisco
5. Jakobi N (1998) Running across the reality gap: Octopod locomotion evolved in a minimal simulation. In: Husbands P, Meyer J A (eds) Evolutionary Robotics: First European Workshop. EvoRobot98. Springer-Verlag
6. Kawai K, Ishiguro A, Eggenberger P (2001) Incremental evolution of neurocontrollers with a diffusion-reaction mechanism of neuromodulators. In: Proceedings of the 2001 IEEE/RSJ International Conference on Intelligent Robots and Systems (IROS'01), vol. 4. Maui, HI
7. Hornby G S, Takamura S, Yokono J, Hanagata O, Fujita M, Pollack J (2000) Evolution of controllers from a high-level simulator to a high dof robot. In: Miller J (ed) Evolvable Systems: from Biology to Hardware; Proceedings of the Third International Conference (ICES 2000). Lecture Notes in Computer Science, vol. 1801. Springer
8. Lipson H, Pollack J B (2000) Automatic design and manufacture of robotic lifeforms. Nature 406(6799):974–978

9. Hoffmann F, Zagal Montealegre J C S (2001) Evolution of a tactile wall-following behavior in real time. In: The 6th Online World Conference on Soft Computing in Industrial Applications (WSC6)
10. Schultz A C, Grefenstette J J, Adams W (1996) RoboShepherd: learning a complex behavior. In: Robotics and Manufacturing: Recent Trends in Research and Applications, vol. 6
11. Keymeulen D, Iwata M, Kuniyoshi Y, Higuchi T (1998) Online evolution for a self-adapting robotic navigation system using evolvable hardware. Artificial Life 4(4):359–393
12. Quinn M, Smith L, Mayley G, Husbands P (2002) Evolving team behaviour for real robots. In: EPSRC/BBSRC International Workshop on Biologically-Inspired Robotics: The Legacy of W. Grey Walter (WGW '02). HP Bristol Labs, U.K.
13. Nolfi S, Floreano D (1998) Co-evolving predator and prey robots: Do 'arms races' arise in artificial evolution? Artificial Life 4(4):311–335
14. Buason G, Bergfeldt N, Ziemke T (2005) Brains, bodies, and beyond: competitive co-evolution of robot controllers, morphologies and environments. Genetic Programming and Evolvable Machines 6(1):25–51
15. Cliff D, Miller G F (1996) Co-evolution of pursuit and evasion II: simulation methods and results. In: Maes P, Mataric M, Meyer J-A, Pollack J, Wilson S W (eds) From Animals to Animats 4: Proceedings of the Fourth International Conference on Simulation of Adaptive Behavior (SAB96). MIT Press/Bradford Books
16. Chellapilla K, Fogel D B (2001) Evolving an expert checkers playing program without using human expertise. IEEE Transactions on Evolutionary Computation 5(4):422–428
17. Lubberts A, Miikkulainen R (2001) Co-evolving a go-playing neural network. In: Coevolution: Turning Algorithms upon Themselves, Birds-of-a-Feather Workshop, Genetic and Evolutionary Computation Conference (GECCO-2001). San Francisco
18. Nolfi S, Floreano D (2000) Evolutionary robotics: the biology, intelligence, and technology of self-organizing machines. MIT Press, Cambridge, MA
19. Harvey I, Husbands P, Cliff D, Thompson A, Jakobi N (1997) Evolutionary robotics: the Sussex approach. Robotics and Autonomous Systems 20(2–4):205–224
20. Meeden L A, Kumar D (1998) Trends in evolutionary robotics. In: Jain L C, Fukuda T (eds) Soft Computing for Intelligent Robotic Systems. Physica-Verlag, New York
21. Harvey I, Husbands P, Cliff D (1994) Seeing the light: artificial evolution, real vision. In: Cliff D, Husbands P, Meyer J-A, Wilson S (eds) From Animals to Animates 3. Proc. of 3rd Intl. Conf. on Simulation of Adaptive Behavior (SAB94). MIT Press/Bradford Books, Cambridge, MA
22. Watson R A, Ficici S G, Pollack J B (2002) Embodied evolution: distributing an evolutionary algorithm in a population of robots. Robotics and Autonomous Systems 39(1):1–18
23. Kodjabachian J, Meyer J-A (1998) Evolution and development of neural networks controlling locomotion, gradient-following, and obstacle avoidance in artificial insects. IEEE Transaction on Neural Networks 9(5):796–812

24. Floreano D, Mondada F (1996) Evolution of homing navigation in a real mobile robot. IEEE Transactions on Systems, Man, Cybernetics Part B: Cybernetics 26(3):396–407
25. Beer R D, Gallagher J C (1992) Evolving dynamical neural networks for adaptive behavior. Adaptive Behavior 1(1):91–122
26. Grefenstette J, Schultz A (1994) An evolutionary approach to learning in robots. Machine Learning Workshop on Robot Learning. New Brunswick
27. Augustsson P, Wolff K, Nordin P (2002) Creation of a learning, flying robot by means of evolution. In: Proceedings of the Genetic and Evolutionary Computation Conference (GECCO 2002). New York
28. Brooks R A (1992) Artificial life and real robots. In: Varela F J, Bourgine P (eds) Toward a Practice of Autonomous Systems: Proceedings of the First European Conference on Artificial Life. MIT Press/Bradford Books, Cambridge, MA
29. Brooks R A (1990) Elephants don't play chess. Robotics and Autonomous Systems 6:3–15
30. Hornby G S, Lipson H, Pollack J B (2001) Evolution of generative design systems for modular physical robots. In: Proceedings of the IEEE International Conference on Robotics and Automation (ICRA'01), vol. 4
31. Capi G, Doya K (2005) Evolution of recurrent neural controllers using an extended parallel genetic algorithm. Robotics and Autonomous Systems 52(2–3):148–159
32. Nelson A L, Grant E, Barlow G J, White M (2003) Evolution of autonomous robot behaviors using relative competitive fitness. In: Proceedings of the 2003 IEEE International Conference on Integration of Knowledge Intensive Multi-Agent Systems (KIMAS'03). Modeling, Exploration, and Engineering Systems. Boston, MA
33. Koza J R (1992) Evolution of subsumption using genetic programming. In: Varela F J, Bourgine P (eds) Toward a Practice of Autonomous Systems: Proceedings of the First European Conference on Artificial Life. MIT Press/Bradford Books, Cambridge, MA
34. Ishiguro A, Tokura S, Kondo T, Uchikawa Y (1999) Reduction of the gap between simulated and real environments in evolutionary robotics: a dynamically-rearranging neural network approach. In: Proceedings of the 1999 IEEE International Conference on Systems, Man, and Cybernetics, vol. 3
35. Lee W, Hallam J, Lund H H (1996) A hybrid GP/GA approach for co-evolving controllers and robot bodies to achieve fitness-specified task. In: Proceedings of IEEE 3rd International Conference on Evolutionary Computation
36. Cliff D, Miller G F (1995) Tracking the red queen: measurements of adaptive progress in co-evolutionary simulations. In: Moran F, Moreno A, Merelo J J, Cachon P (eds) Proceedings of the Third European Conference on Artificial Life: Advances in Artificial Life (ECAL95). Lecture Notes in Artificial Intelligence 929. Springer-Verlag
37. Nolfi S (1997) Evolving non-trivial behaviors on real robots. Robotics and Autonomous Systems 22(3–4):187–198
38. Ziemke T (1999) Remembering how to behave: recurrent neural networks for adaptive robot behavior. In: Medsker, Jain (eds), Recurrent Neural Networks: Design and Applications. CRC Press, Boca Raton
39. Floreano D, Urzelai J (2000) Evolutionary robots with on-line self-organization and behavioral fitness. Neural Networks 13(4–5):431–443

40. Tuci E, Quinn M, Harvey I (2002) Evolving fixed-weight networks for learning robots. In: Proceedings of the 2002 Congress on Evolutionary Computing, vol. 2. Honolulu, HI
41. Ashiru I, Czarnecki C A (1998) Evolving communicating controllers for multiple mobile robot systems. In: Proceedings of the 1998 IEEE International Conference on Robotics and Automation, vol. 4
42. Quinn M (2000) Evolving cooperative homogeneous multi-robot teams. In: Proceedings of the IEEE/RSJ International Conference on Intelligent Robots and Systems (IROS'00), vol. 3. Takamatsu, Japan
43. Baldassarre G, Nolfi S, Parisi D (2002) Evolving mobile robots able to display collective behaviors. In: Hemelrijk C K, Bonabeau E (eds) Proceedings of the International Workshop on Self-Organisation and Evolution of Social Behaviour. Monte Verit, Ascona, Switzerland
44. Dorigo M, Trianni V, Sahin E, Labella T, Grossy R, Baldassarre G, Nolfi S, Deneubourg J-L, Mondada F, Floreano D, Gambardella L (2004) Evolving self-organizing behaviors for a swarm-bot. Autonomous Robots 17(23):223–245
45. Nordin P, Banzhaf W, Brameier M (1998) Evolution of a world model for a miniature robot using genetic programming. Robotics and Autonomous Systems 25(1–2):105-116
46. Nelson A L, Grant E, Barlow G J, Henderson T C (2003) A colony of robots using vision sensing and evolved neural controllers. In: Proceedings of the 2003 IEEE/RSJ International Conference on Intelligent Robots and Systems (IROS'03). Las Vegas NV
47. Gomez F, Miikkulainen R (2004) Transfer of neuroevolved controllers in unstable domains. In: Proceedings of the Genetic and Evolutionary Computation Conference (GECCO-04). Seattle, WA
48. Earon E J P, Barfoot T D, D'Eleuterio G M T (2000) From the sea to the sidewalk: the evolution of hexapod walking gaits by a genetic algorithm. In: Proceedings of the International Conference on Evolvable Systems (ICES). Edinburgh, Scotland
49. Zykov V, Bongard J, Lipson H (2004) Evolving dynamic gaits on a physical robot. In: 2004 Genetic and Evolutionary Computation Conference (GECCO). Seattle, WA
50. Chernova S, Veloso M (2004) An evolutionary approach to gait learning for four-legged robots. In: Proceedings of the IEEE International Conference on Intelligent Robots and Systems (IROS'04), vol. 3. Sendai, Japan
51. Zufferey J, Floreano D, van Leeuwen M, Merenda T (2002) Evolving vision based flying robot. In: Blthoff, Lee, Poggio, Wallraven (eds) Proceedings of the 2nd International Workshop on Biologically Motivated Computer Vision LNCS 2525. Springer-Verlag, Berlin
52. Macinnes I, Di Paolo E (2004) Crawling out of the simulation: evolving real robot morphologies using cheap, reusable modules. In: Proceedings of the International Conference on Artificial Life (ALIFE9). MIT Press, Cambridge, MA
53. Nakamura H, Ishiguro A, Uchilkawa Y (2000) Evolutionary construction of behavior arbitration mechanisms based on dynamically-rearranging neural networks. In: Proceedings of the 2000 Congress on Evolutionary Computation, vol. 1

5

Evolving Fuzzy Classifier for Novelty Detection and Landmark Recognition by Mobile Robots

Plamen Angelov, Xiaowei Zhou

Communication Systems, Infolab21,
Lancaster University, LA1 4WA, UK
{p.angelov, x.zhou3}@lancaster.ac.uk,
http://www.lancs.ac.uk/staff/angelov/

In this chapter, an approach to real-time landmark recognition and simultaneous classifier design for mobile robotics is introduced. The approach is based on the recently developed evolving fuzzy systems (EFS) method [1], which is based on subtractive clustering method [2] and its on-line evolving extension called eClustering [1]. When the robot travels in an unknown environment, the landmarks are automatically deteced and labelled by the EFS-based self-organizing classifier (eClass) in real-time. It makes fully autonomous and unsupervised joint landmark detection and recognition without using the absolute coordinates (altitude or longitude), without a communication link or any pre-training. The proposed algorithm is recursive, non-iterative, incremental and thus computationally light and suitable for real-time applications. Experiments carried out in an indoor environment (an office located at InfoLab21, Lancaster University, Lancaster, UK) using a Pioneer3 DX mobile robotic platform equipped with sonar and motion sensors are introduced as a case study. Several ways to use the algorithm are suggested. Further investigations will be directed towards development of a cooperative scheme, tests in a realistic outdoor environment, and in the presence of moving obstacles.

5.1 Introduction

Classification has been applied to pattern recognition problems including landmark detection for quite a few years [6],[22]. Most of the approaches however, process the data off-line in a batch mode to generate and train the classifier. Afterwards, the trained model can be applied to new data inputs fed online. The underlying assumption is that the statistical characteristics of the validation data set are similar to that of the training data set, so that the validation can be made on different data set in off-line mode. The disadvantage of such a

P. Angelov and X. Zhou: *Evolving Fuzzy Classifier for Novelty Detection and Landmark Recognition by Mobile Robots*, Studies in Computational Intelligence (SCI) **50**, 89–114 (2007)
www.springerlink.com

scheme is that changes in the data pattern or unexpected data are not taken into account by an off-line trained classifier [6].

In the task of landmark recognition, however, it is vitally important to have the ability to adapt the classifier to new data patterns so the robots can work in the unpredictable and changing environment.

Some adaptive classifiers have been reported in the literature that are based on estimation using Bayes' theory [7] or the heuristic random search using genetic/evolutionary algorithms [8]. The adaptation in the stochastic classifiers [7], however, concern only the statistical properties of the data and not the structure of the classifier. The evolutionary classifiers [8], however, are computationally expensive, and operate over a population of candidate solutions applying so-called 'crossover', 'mutation' operations and 'reproduction', thus are prohibitive for real-time applications. In addition, both groups of approaches include supervision. It is well known that classification, by definition, assumes supervision/labeling of the classes [6], which is a serious obstacle in designing autonomous and flexible adaptive systems.

Alternatively one can use the self-organizing maps (SOM), introduced originally by T. Kohonen in 1987 for unsupervised learning (clustering)[13]. They are computationally less expensive and have been developed further into evolving SOM (eSOM) with clusters that 'evolve' [14]. However, eSOM, as well as a number of other evolving and self-organizing neural networks such as growing cell structures [15], adaptive resonance theory [16], generalized growing and pruning radial-basis function networks [17], evolving fuzzy neural networks [11], dynamic evolving neuro-fuzzy inference systems [18], resource allocation networks [19] do not take data density into account. The result is that they tend to generate too many clusters, some of which usually have negative effect on the performance. Thus, pruning is necessary which reduces the quality of the fuzzy rule-based classifier. All these approaches are not prototype-based in the data space. Instead of locating on real points, the centre of the clusters are usually located at the points, such as the mean or the points that result from an adaption. Additional disadvantage of these approaches is that new data point is compared to the cluster centers only, as the computing resource are limited for real-time applications which precludes memorizing the data history.

A new approach to real-time data clustering was proposed recently [1] which is developed from the well known subtractive clustering [2] and Mountain clustering approaches [3]. This method is fully unsupervised in the sense that the number of clusters is also not pre-defined but determined based on the data spatial distribution in the feature space. This approach [1],which is called evolving Clustering (eClustering), has been used for real-time data partitioning and was combined with an extended version of the recursive least squares estimation for real-time generation of Takagi-Sugeno type fuzzy rules from data [4].

In this chapter, the concept of data density measured by its "*potential*" used in subtractive and eClustering is applied as a basis for landmark recognition in mobile robotics. An alternative formula for *potential* (data spatial proximity) calculation is introduced in the chapter that is using the concept of participatory learning [20]. The data points with low *potential* are treated as 'landmarks' in exploring a new environment by a mobile robot due to their specific/outstanding position in the feature space. In the experiment settings, only on-board sensors and computational device are used, with no pre-training, pre-installed knowledge, nor external communication or externally linked device, such as GPS.

Real-time data are classified into automatically labeled classes associated to the landmarks, or into a default class corresponding to the normal routine behavior. The number of classes respectively the landmarks is not prespecified. Instead it starts evolving 'from scratch' with the very first landmark detected while exploring unknown environment. The EFS-based classifier proposed in this paper, namely eClass, is formed by real-time detection and labeling of landmarks. eClass is then used to classify the data produced by the sonar and motion sensors mounted on a mobile robotic platform Pioneer3-DX in real-time. In the experiment carried out in an indoor office environment (office B-69, InfoLab21, South Drive, Lancaster, UK) a robot performed 'wall following' behavior [9] exploring the unknown environment. The landmarks in the empty office are associated with the corners of that office. Each corner differs by its type (convex or concave) and relative position. The results illustrate the superiority of the proposed approach in terms of computational efficiency, precision and flexibility, when comparing with the other approaches applied to similar settings.

Future investigations will be directed towards development of a co-operative scheme, tests in a realistic outdoor environment, and with moving obstacles.

5.2 Landmark Recognition in Mobile Robotics

The sensors installed on a robot, such as sonar, laser, motion controller, etc, generate a stream of data sampled with certain rate. When the robot travels in a previously unseen environment, it collects information about this new environment. This data includes both a 'routine' stream of data and some interesting new patterns. 'Novelty detection', namely the ability to differentiate between existing (common sensory) data patterns and new data patterns is a very useful competence for mobile robots exploring real dynamic environments. With such ability, the robot is able to identify the characteristics of the environment which differ from the contextual background. The specific characteristics of the environment are then used as 'landmarks'.

In self-localization, adaptive navigation and route planning applications of autonomous robots, landmarks are used as the key reference points when performing tasks in the unknown environment. Therefore, the ability to automatically detect, label, identify and use landmarks based on the environment is vital.

At the same time, the computational resources available to autonomous mobile robots are often limited, due to the cost and size of the robots. As large amount of online sensory data are required to be processed in real-time, a recursive algorithms are highly desirable that are able to cope with the challenge of limited memory and time. As there are growing demand for agile compact autonomous devices, this becomes more critical due to restrictive requirements on computation and energy.

'eClass', the evolving fuzzy system-based classifier that is described in this chapter, is an efficient solution that is addressing the problem of real-time novelty detection and landmark recognition, and simultaneous classifier generation. It requires very low computational resource in terms of both memory and processing time due to its recursive, one pass, and non-iterative feature. On the other hand, the algorithm requires no pre-training. No human intervention is required, which enables the learning procedure to starts 'from scratch' fully automatic. These features of eClass make it suitable for real-time applications.

5.3 Evolving Fuzzy Rule-Based Classifier (eClass)

In the following section, the suggested evolving classifier, eClass is introduced in detail, from its mechanism of classification and evolution, to the structure and procedure in application to landmark recognition. The summarized pseudo code of eClass is also given in later part of the section.

As the name suggests, the formation and number of the classes generated by eClass are not pre-defined and gradually evolve in real time with data input in time series. Class labels are automatically assigned with alphabetic symbols('A','B','C'...) when a class is generated. eClass starts with a default class labelled '@' corresponding to the normal routine behavior and empty rule-base. The routine behaviors are usually motions of the robot to keep straight headings and constant distance to the reference objects defined by the route planning algorithm, such as object-follow, wall-follow, space-follow, etc. In the algorithm used for landmark recognition, all the data are described as the routine/normal behavior unless a landmark is detected.

The data is read incrementally sample by sample in real-time. Each data sample is described by a vector representing a data point in the multi-dimensional feature space, $x_k = \left[x_k^1, x_k^2, ..., x_k^n\right]^T$, data in each dimension represents readings from one sensor device, where k is the current time instant. Normalization is required on each dimension to set data range between $[0, 1]$, which ensures that the information from different devices will be equally treated. For the mobile robotic application considered in this chapter, the n-dimensional data vector includes the sensor readings that are available at given time instant k. For example, rotation, and the distance to the nearest obstacle, etc. The time instant k (k=1,2,3,...) is used to represent the current moment of time. In a real-time application, the time series is open-ended. The algorithm processes the inputs and then gets the next inputs and stops only when an external stop condition is reached.

5.3.1 The Informative Data Density and Proximity Measure

eClass, similarly to eClustering, is based on the concept of '*potential*', which represents a data spatial density measurements [1]. This concept originates from the mountain function [3] which later has been modified into *potential* in the subtractive clustering method [2]. Gaussian exponential is used to describe the *potential* in subtractive clustering while a computationally simpler Cauchy function and the so-called scatter are used in [5]. An alternative formula that combines the concept of participatory learning, proposed by Yager [3], and the concept of scatter can be given by:

$$P_k = P(x_k) = 1 - \frac{\sum_{i=1}^{k-1} \|x_i - x_k\|^2}{n(k-1)} \tag{5.1}$$

where P_k denotes the *potential* of the kth data point x_k; $\|.\|$ denotes the Euclidean distance; n denotes the number of dimensions.

The *potential* formulated in this way represents a measurement of the compatibility of the information brought in by new inputs compared with the existing information. The measurement is mapped into the data space as inverse of the normalized accumulated Euclidean distances between new sensor readings and all previous sensor readings, which included density information in the data space. Taking spatial density-related information in to account makes the difference between eClass/eClustering and other classification/clustering approaches used for data space partitioning in self-organized neuro-fuzzy models.

Please note that the original form of *potential* calculation 5.1 is only suitable for off-line (batch) calculation as it requires memorizing *all* previous sensor readings for summation to determine the density. In order to be accommodated in a real-time algorithm, a recursive version of 5.1 which avoids memorizing the data history is demanded. The explicit form of the projections of the squared distances on axes $[0; x^j)$ from 5.1 can be derived as:

$$P_k = 1 - \frac{\sum_{i=1}^{k-1} \sum_{j=1}^{n} \left\{ \left(x_k^j\right)^2 - 2x_k^j x_i^j + \left(x_i^j\right)^2 \right\}}{n(k-1)} \tag{5.2}$$

By reorganizing 5.2 we get:

$$P_k = 1 - \frac{(k-1) \sum_{j=1}^{n} \left(x_k^j\right)^2 - 2 \sum_{j=1}^{n} x_k^j \sum_{i=1}^{k-1} x_i^j + \sum_{i=1}^{k-1} \sum_{j=1}^{n} \left(x_i^j\right)^2}{n(k-1)} \tag{5.3}$$

Assume the following notations are used:

$$a_k = \sum_{j=1}^{n} (x_k^j)^2; b_k = \sum_{i=1}^{k-1} \sum_{j=1}^{n} (x_i^j)^2; c_k = \sum_{j=1}^{n} x_k^j f_k^j; f_k^j = \sum_{i=1}^{k-1} x_i^j \tag{5.4}$$

The original *potential* formula is transformed into the recursive formula:

$$P_k = 1 - \frac{a_k(k-1) - 2c_k + b_k}{n(k-1)} \tag{5.5}$$

Values a_k and c_k represent accumulated projections on each dimension of the cartesian data space $[0; x^j)$ and can be calculated only requiring the current data point x^k (current sensor readings) are available. The values b_k and f_k^j require accumulation of past information. To recursively update the accumulation, two auxiliary variables scaler b_{k-1} and vector column $f_k = \left[f_k^1, f_k^2, ..., f_k^n\right]^T$ are used:

$$b_k = b_{k-1} + a_{k-1}; b_0 = 0 \tag{5.6}$$

$$f_k^j = f_{k-1}^j + x_{k-1}^j; f_0^j = 0 \tag{5.7}$$

Then, using 5.4-5.7, the *potential* (data spatial density) can be calculated by using only the current data sample with the two recursively updated auxiliary variables (b_k and f_k^j), therefore, it is unnecessary to store the historical data any more. This recursively calculated *potential* keeps the information of spatial data density regarding the whole previous history in the data space without storing the history of the data in the memory. Mean while, it also avoids the computation time for processing the big and growing amount of history data, which means a short response time in real-time. This feature makes the proposed algorithm efficient (computationally light) in both speed and storage and enables it to be a strong candidate for real-time applications.

5.3.2 Landmark Classifier Generation and Evolution

When a robot travels into a new environment, the combination of the readings from selected sensor devices are different to those which are generated from the known environment or when performing the routine behaviors. Ideally, when the vector containing these readings is plotted in the data space, the accumulated distance to all existing data points will be large, which means the point is far away from the existing points. Therefore, according to formula 5.1, the *potential* of this newcoming point will be low.

The data points with distinctively low *potential* can be a candidate representative of a landmark. Thus, in the algorithm, low value of *potential* indicates a possible demand to introduce a new landmark:

$$P_k < \underline{P} \tag{5.8}$$

where $\underline{P}$ is a positive threshold. If the value of this threshold is set too high, the condition to be "distinctively low" become looser, therefore, too many landmarks will be generated, some of them are actually routine behavior; contrarily, if the value of the threshold is too low, the condition become more restrictive, and consequently, some landmarks can be incorrectly classified into routine behavior. Instead of arbitrarily set the threshold, it is possible to use the minimum potential of focal points (prototype class centre) of the classes as the threshold.

In the process cycle of the algorithm, the very first data point that satisfies equation 5.8 is assumed to be the first landmark and is automatically assigned with label 'A':

$$x_{*1} \leftarrow x_k \ when \ P_k < \underline{P} \tag{5.9}$$

A fuzzy rule is generated when the class is formed. The antecedent part of the fuzzy rule is formed around the prototype class centre, and the consequence is defined as crisp (non-fuzzy) class:

$$\begin{aligned} R_1: \ & IF \ (x_k^1 \ is \ around \ x_{*1}^1, \ AND \ x_k^2 \ is \ around \ x_{*1}^2, \\ & \ldots, \ AND \ x_k^n \ is \ around \ x_{*1}^n \\ & THEN \ (Class \ is \ 'A') \end{aligned} \tag{5.10}$$

Instead of using the points get from calculation, such as mean of supporting points, we use real point (prototype) as class centre. The reason is that, it may not be possible for the points from calculation to exist in the real world, and then, the corresponding rules will lose linguistical interpretability.

The algorithm continues reading the next data point and calculating its *potential*. New classes are formed based on verifying the potential.

A situation may occur when several landmark candidates are with low *potential*, some of them are very close to the existing class centers in the data space. Ambiguity should be avoided in the selection of class representing a landmark. There should be no indistinguishable, ambiguous landmarks, so called 'perceptual aliasing' [12]. Therefore, additional condition that candidate landmarks should not be in the vicinity of any existing class centers is introduced in addition to the threshold formula 5.8:

$$\|x_{*i} - x_k\| > r/2; \ i = 1, 2, \ldots, N; k = 1, 2, \ldots \tag{5.11}$$

where x_{*i} denotes i^{th} class center (landmark); N is the total number of classes (landmark). The parameter r is a threshold that determines the zone of influence of a specific landmark centered at the class focal point.

During the stage of landmark candidate verification to the new data point, both conditions 5.8 and 5.11 are tested. If both conditions are satisfied, which means the new data point is with low potential and not close to any existing landmark, a new landmark is detected and a new class is labeled with the next symbol(character), this data point becomes the center of the class:

$$x_{*i+1} \leftarrow x_k \ when \ \|x_{*i} - x_k\| > r/2 \ and \ P_k < \underline{P} \tag{5.12}$$

If condition 5.8 is satisfied while condition 5.11 is violated, which means the new data point with low potential is close enough to an existing class center, it replaces the nearest landmark:

$$x_{*i} \leftarrow x_k \; when \; \|x_{*i} - x_k\| \leq r/2 \; and \; P_k < \underline{P} \tag{5.13}$$

In this way, an EFS-based classifier is generated in real-time with a recursive, non-iterative, incremental (one pass) algorithm that is based on the data spatial density (*potential*).

5.3.3 Landmark Recognition (real-time classification)

At each time instant, when the inputs sensor readings are processed, classification for landmark recognition and the generation and evolution of classifier are taking place in the same cycle simultaneously.

For the particular application of exploring an unknown environment by a mobile robot, two general states are usually defined:

1. Normal routine operation,including 'wall following', 'random walk' or following certain navigation goal,etc. These behavior corresponds to the default class '@';
2. Novelty detection and landmark recognition, namely, classifying the data into one of the existing classes: $ClassA$, ..., $ClassZ$, or create a new Class: $ClassAA$. New class may also replace an existing class.

The data fed from the normal routine operation, may drift in a small range, however, as the characters of the routine behavior, the data pattern is largely the same. At the very early stage, when the first distinct data point is detected with low potential, the first landmark is therefore generated. Following this time instant, further data falls into the case 1 or 2 described above. The task of the real-time classification is to classify the incoming data into general class 1 or 2. In case of 2, new data point needs to be assigned to particular classes presenting different landmarks, or a new class needs to be generated as a new landmark.

The algorithm always tries to classify the sensor readings into one of the classes in case 2. The data point will be classified into case 1 only when the conditions set for class 2 have failed. In the algorithm, the default case for routine behavior is given with an ELSE branch in the program flow:

A landmark can also replace a previously existing landmark if condition 5.13 is satisfied. The default case (normal/routine behavior) is given with an ELSE construct:

$$\begin{aligned} R_1: \; & IF\ (x_k^1\ is\ around\ x_{*1}^1,\ AND\ x_k^2\ is\ around\ x_{*1}^2, \\ & \ldots,\ AND\ x_k^n\ is\ around\ x_{*1}^n \\ & THEN\ (Class\ is\ 'A')\ ELSE\ (Class\ is\ '@') \end{aligned} \tag{5.14}$$

The process starts once the first class, respectively the first fuzzy rule, has been generated. Before the initial formation of the rule base, all data points are classified into Class '@'.

eClass is a fuzzy rule-base system, the de-fuzzification operation for overall classification used is based on the 'winner take all' principle [3], which corresponds to the MAX t-co-norm. The same de-fuzzification operator is also used in Mamdani type fuzzy models and therefore the fuzzy rule-based classifier eClass can be considered either as zero order Takagi-Sugeno type or as simplified Mamdani type (because the consequent part is crisp, consisting of non-fuzzy singletons) [9]:

$$y = y_{win};\ win = \arg\max_{j=1}^{N}(\lambda_j) \tag{5.15}$$

where y_{win} represents the winner class (landmark) $\lambda_i^j = \begin{cases} 1 & \sum_{i=1}^{N} \tau_i = 0 \\ \frac{\tau_j}{\sum_{i=1}^{N} \tau_i} & otherwise \end{cases}$ is the normalized firing level of the j_{th} fuzzy rule, $j = 1, 2, ...N$. The activation level, can be defined as a Cartesian product or conjunction (t-norm) of respective fuzzy sets for this rule [3]:

$$\tau_i = \mu_i^1(x^1) \times \mu_i^2(x^2) \times ... \times \mu_i^n(x^n) \tag{5.16}$$

where $\mu_i^j$is the membership value of the j_{th} input $x^j (j = 1, 2, \ldots, n)$ to the respective fuzzy set for the i_{th} fuzzy rule ($i = 1, 2, \ldots, N$)

The return value of the membership function at specific data point indicates the degree that the data point 'belongs' to the fuzzy set in respect to the similarity to the landmark. Triangular membership function is applied in the experiments introduced in later case study:

$$\mu_i^j(x_k) = \begin{cases} 1 - \frac{|x_k^j - x_{*i}^j|}{r_i^j} & x_k^j - x_{*i}^j < r_i^j \\ 0 & otherwise \end{cases} \tag{5.17}$$

where r_i^j is the radius of the zone of influence of the i_{th} landmark in its j_{th} dimension.

One can also use a Gaussian type membership function (bell-shaped, figure 5.1):

$$\mu_i^j(x_k) = e^{-\frac{|x_k^j - x_{i*}^j|}{2r_i^j}} \tag{5.18}$$

As data from different sensor devices may have different importance in real application, radius can be different in each dimension in the data space correspond to different importance or weight of reading from each sensor. Also,

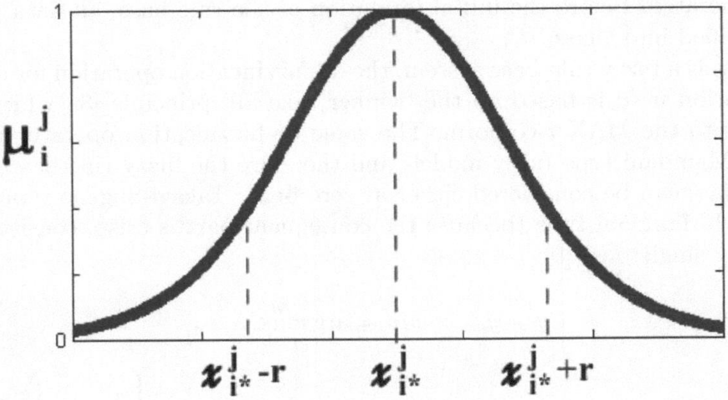

Fig. 5.1. Gaussian type membership function

for different landmark in the environment, the level of feature distinction are usually different even in the same data space. The 'online adaptive radius' has been recently introduced into eClustering algorithm [10] and similar idea can be applied into eClass in certain circumstance.

If analyzing 5.17 or 5.18, one can conclude that all of the membership functions describing closeness to a landmark will have value zero for the routine behavior ($\mu_i^j(x_k) = 0, \;\Rightarrow\; \sum_{i=1}^{N} \tau_i = 0, \;\Rightarrow\; \lambda = 1 \; for \; \forall j = 1,2,\ldots,n$ $and \; i = 1,2,\ldots,N$) .

It is interesting to note that due to the strong condition of 5.8, which concerns the density information from previously data in the data space, the frequency of generating new rules and the number of fuzzy rules representing distinct landmarks does not grow excessively. In addition, the mechanism to modification of existing cluster center by replacing centers with qualified data points also prevents enlarging the rule-base excesively.

5.3.4 Learning of eClass

The learning procedure for eClass can be divided into 3 stages. First initialization is done at the beginning, outside the online data process cycle. The classifier starts with an empty fuzzy rule base. The first data point (a n-dimensional vector of normalized sensor readings at an instant of time) is read in real-time. Its *potential* (data spatial density) is assumed to be equal to 1, $P(x_1) \leftarrow 1$ and the data point is assigned to Class '@' (routine behavior).

Then the processing loop starts from the second data point. With the next data point, the *potential* is calculated recursively using 5.5. Two auxiliary variables (the one dimensional scalar b and the n-dimensional vector f) are accumulated according to 5.4-5.7. Once a data point has been used it is *discarded* and not stored in the memory, which allows computational efficiency and real-time application.

During the routine behavior, no landmarks were detected and no classes were formed, before the first corner is met. Thus the winning class is the default (Class '@'). The first data point that satisfies equation 5.8 is used to form the first class:

$$x_{*1} \leftarrow x_k; \; N \leftarrow 1 \tag{5.19}$$

Label 'A' is assigned to this class and a fuzzy rule of type 5.14 is generated. For each new data point:

1. its *potential* is calculated recursively using 5.4-5.7;
2. conditions 5.8 and 5.11 are checked;
3. based on the verification of the data point, one of the following actions are taken:
   a) **IF** (5.13 holds) **THEN** (replace a cluster center that is closer to the new data point $x_{*i} \leftarrow x_k$);
   b) **ELSEIF** (5.12 holds) **THEN** (form a new cluster around the new data point ($x_{*N+1} \leftarrow x_k$); assign a new label $N \leftarrow N+1$; form a new fuzzy rule of type 5.10 - 5.14);
   c) **ELSEIF** ($\delta_r < r$; $\delta_1 = \|x_{i*} - x_k\|$; $\gamma = \arg\min_{i=1}^{N}(\|x_{i*} - x_k\|)$) **THEN** (assign the new data point to the Class)
   d) **ELSE** (classify the behavior as routine (assign the data point to Class 0 and do not change the classifier structure).

The loop continues with the input data being classified repeatedly, till no more data is available or an external instruction to stop the process is received. The formal Procedure can be summarized with the following pseudo-code:

5.4 Case Study: Corner Recognition

In this section, we introduce an experiment which illustrates the application of the proposed algorithm eClass for a simple landmark recognition task. The experiment is carried out by a fully independent mobile robot in an empty office enclosure with corners. The robot comes with no previous knowledge of the environment, and nor external support such as global coordinates or communication links are available. Sonar and motion controllers are used for the robot to perform a "wall-following" operation. Rotation information from the motion controller and reading from one sonar at the back of the robot are feeded as inputs in real-time to the classifier, which is set to identify the corners in the enclosure.

The experiment can be expanded to more complicated real applications with more informative inputs to eClass model from different number and different type of sensory devices.

Algorithm 5.1 Evolving Classification (eClass) for Landmark Recognition

Input: real-time sensory data vector at time k, x_k
Output: on-line assigned landmark label, $ClassNumber$

Begin
  Initialize:
  Read x_1;
  Set $k \leftarrow 1$; $N \leftarrow 1$; $P(x_1) \leftarrow 1$; $b_1 \leftarrow 0$; $f_1^j \leftarrow 0$;
  Do for $k \leftarrow k+1$
    Read $x_k + 1$;
    Calculate P_k recursively using 5.4-5.7;
    Accumulate values b and f using 5.6-5.7;
    Compare P_k with $\underline{P}$ and condition 5.11;
    IF (5.13 holds)
    THEN (replace a cluster centre that is closer to the
      new data point $x_i \leftarrow x_k$);
    ELSEIF (5.12 holds)
    THEN (**form** a new cluster around the new data point $x_i \leftarrow x_k$;
      assign a new label $N \leftarrow N+1$; form a new fuzzy rule of type 5.10 - 5.14)
    ELSEIF ($\delta_r < r$; $\delta_1 = \|x_{i*} - x_k\|$; $\gamma = \arg\min_{i=1}^{N}(\|x_{i*} - x_k\|)$)
    THEN (**assign** x_k to the Class γ);
    ELSE (**classify** the behavior as routine (Class @) and do not
      change the classifier structure);
    END;
    assign the data sample to the nearest class with winner takes all strategy;
  END DO;
END (eClass)

5.4.1 The Mobile Robotic Platform

The autonomous mobile robot Pioneer-3DX [21], supplied by ActivMedia, USA, (figure 5.2) has an on-board controller, an onboard computer (Pentium III CPU, 256 MB RAM), camera, digital compass, sonar and bumper sensors, wireless connection for transmission of status data to a desktop or laptop in real time. (ActivMedia and Pioneer Robots are the registered trademark of ActivMedia Robotics.) The robot can be controlled from the on-board computer through the embedded controller ARCOS [21] in a client-server mode. Pre-programmed behaviors, such as 'random wandering', 'obstacle avoidance', etc can be loaded to the onboard computer.

The fully autonomous behavior of 'detection novelties and landmark recognition' was realized with the eClass for classification and ARIA class library for robot control. ARIA is a set of C-based foundation classes for control of the Active Media robots that runs on top of ARCOS [21].

We defined a five-layer architecture for the single robotic system with ActiveMedia Pioneer-3DX robots. (See figure 5.3.)

Fig. 5.2. Pioneer3 DX mobile robot

The lowest layer is where the devices are situated. Optional devices such as sonar, laser, compass, motors, etc, can be mounted onto the robots. Communications for the control and sensory data from these devices are made through the interface to the onboard controller in the "robot server" layer.

The embedded controller in the robot server layer deals with lower level details of the controls to hardware installed on the robot systems. Based on a client-server architecture, the embedded controller works as the server providing transparent control to the clients through the ActivMedia Robot Control and Operations Software (ARCOS).

At the robot client layer, the onboard computer works as a client of the robot server. In this layer, control programs are loaded onto this computer, by which robot behaviors are controlled. In order to simplify the programming to the robot, ActivMedia provided a set of C++ based object-oriented package, ARIA, to enable close but easy access to the robot. Basically, ARIA provides robot users with controls to the *atomic* behaviors of the robot, such as maintain the velocity of the wheels, adjust the robot headings, getting readings from the sensors, setting robot status, etc. Most of the interaction between user program and the robot are implemented using ARIA as an interface and as a tool. Our program for Landmark Recognition experiment is also built upon ARIA package.

More details about ActivMedia ARCOS and ARIA can be found at ActivMedia Robotics' support webpages:

http://www.activrobots.com
http://robots.activmedia.com

On top of the robot client layer, we have "application objects", which are the modules that implements specific tasks such as route planning, mapping, object tracking, object following, etc. Information from different devices of

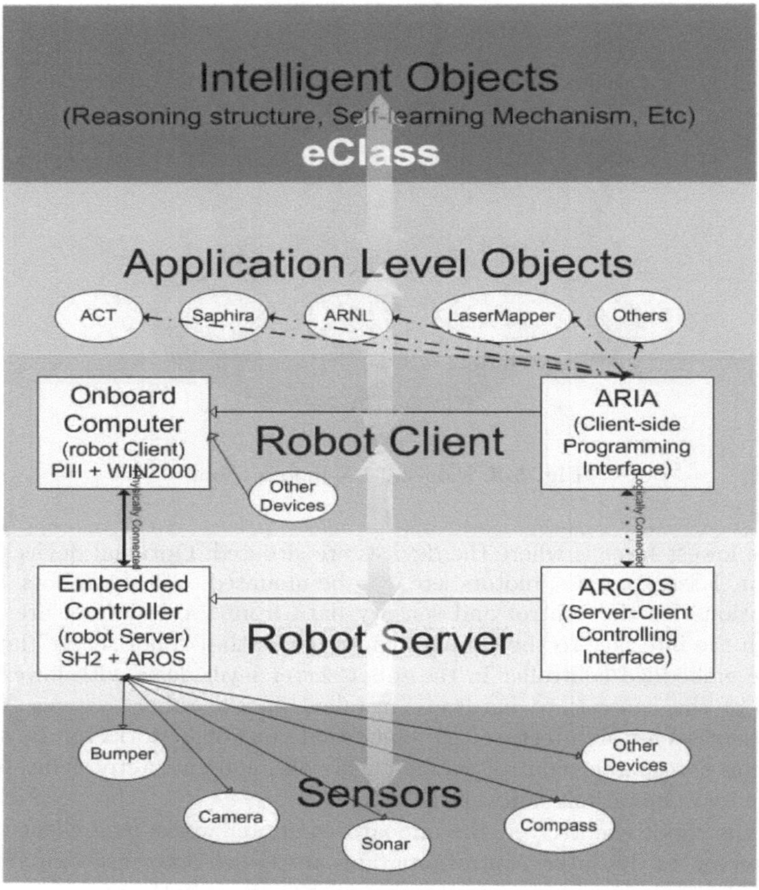

Fig. 5.3. Architecture of the robotic system

the robot are gathered and analyzed together to support the performance of the tasks.

"Intelligent objects" are defined as the mechanism that provides the intelligent solutions to the application models that require reasoning, self-learning, decision making capabilities. Usually, a rule-base is the core of the "intelligent objects". The rule-base is different in different solutions. In our case, an evolvable fuzzy rule base is presented.

Our study will concentrate on the top two layers, especially the Intelligent objects layer, where eClass belongs to. The interface of the robot client layer ARIA is also used in our application.

5.4.2 Experiment Settings

The experiments were conducted in a real indoor environment (an empty office, B-69 located at InfoLab21, Lancaster, UK). In the first step, the similar environment as the one used in [22] was re-created to make possible correct comparison of the results. It comprises of a rectangulary shaped empty office room with 8 corners (6 concave and 2 convex), as sketched in Figure 5.4.

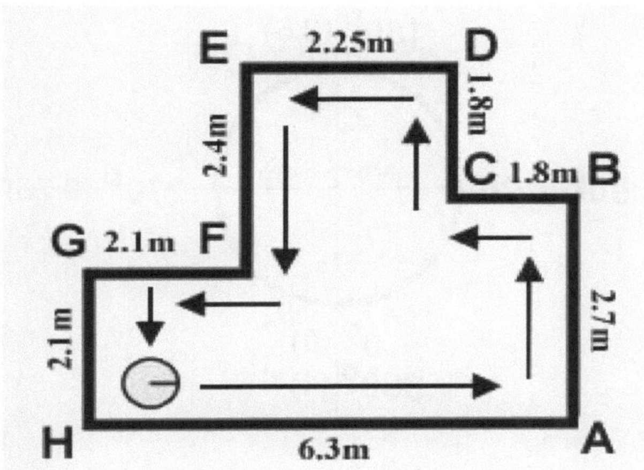

Fig. 5.4. Experimental enclosure

The objective of the experiment is to use a fully autonomous robot to identify all the landmarks, namely the corners in the completely unknown office environment. The task is performed in real-time while the robot is exploring the office with a 'wall following' algorithm as its routine planner. The robot does not have any prior knowledge about the office shape before starting the 'wall following' algorithm. There is not any supervision information, such as global coordinate information from external source during the whole process of the task.

As mentioned earlier, the original idea of the experiment is to compare the performance of eClass in this specific task with the solution presented in the similar experiment using a pre-trained SOM neural network with a fixed structure consisting of 50 neurons. [22]

During the 'wall-following' operation, two sources of information (features) are taken from the robot devices and fed back to eClass as inputs in real-time. One of them is the rotation when the robot adjusts its heading to keep constant distance to the wall including the operation to steer around the corners. It is measured in degrees, normalized by 360° clockwise from the backward direction of the robot heading,and denoted by θ, as in figure 5.5. The other one is the distance to the nearest obstacle detected by the back sonar of the robot, normalized by the range of the sonar M ($M = 15m$ is used in the experiment), denoted by d. In this way, heading straight corresponds to value $\theta = 0.5$; turning $-90°$ (left) in respect to the heading corresponds to $\theta = 0.25$; turning $90°$ (right) in respect to the heading corresponds to $\theta = 0.75$ and turn back corresponds to $\theta = 1$.

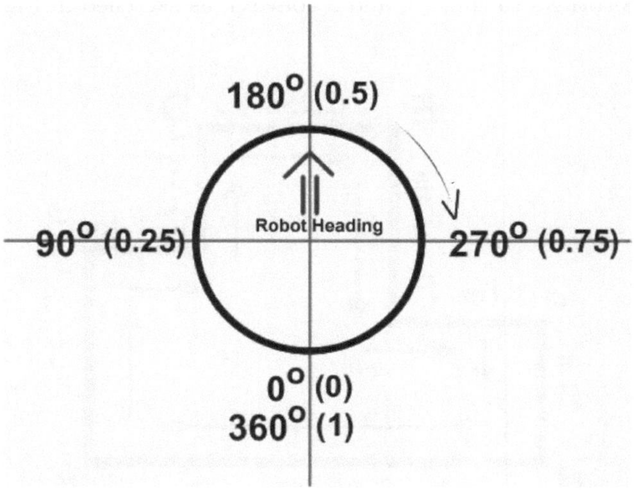

Fig. 5.5. Normalized rotation

Input data are normalized into the range $[0, 1]$. And then, they are organized into an inputs vector (figure 5.6):

$$x = [\theta, d] \tag{5.20}$$

It should be mentioned that the features that are selected for the input vector have critical effects on the performance of the classifier. Experiments were conducted with between 2 and 5 features. Generally, including more features leads to a more refined detection and eventually to a better result, but requires more computing power. A discussion will be given later in this chapter on choosing features as inputs. Finally, each landmark is labeled to represent a distinctive class with alphabetical characters $('A','B','C',\ldots)$ (figure 5.6).Thus, the generated classes are closely related to the corners. Ideally, if we have 8 corners in the office, there should be 8 landmarks and, respectively, 8 classes corresponding to each corner, and the default class '@' corresponding to the routine behavior.

5.4.3 Program Structure

The structure of the experimental program consists of three fundamental component modules (figure 5.7):

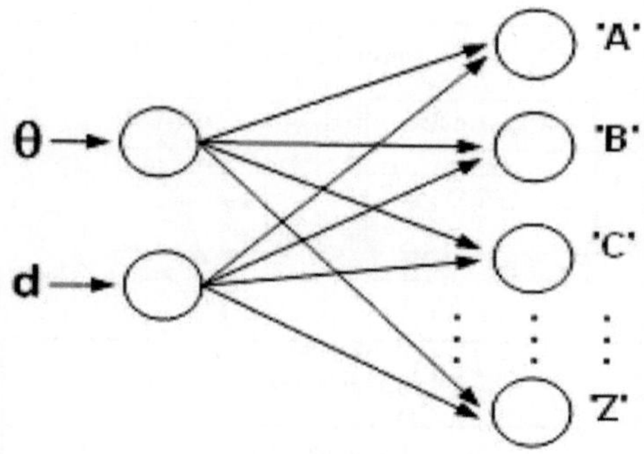

Fig. 5.6. eClass Neuro-Fuzzy System

- Robot controller – control the robot motions and get the readings from the sensor devices.
- Classifier – takes and selects features from the sensor readings, robot motions, and robot status as inputs to the classifier; classifies input data samples; and labels the "landmarks".
- Central controller – controls the wall following behavior of the robot, manages the communication link between the robot control module and the classifier module.

The experimental program is coded in the programming language C++. The ARIA classes ArRobot and ArSonarDevice are applied to compose the robot control module, therefore, no additional codes are necessary for the control of a single (atomic) behavior of the robot and sensors. The 'wall following' algorithm is implemented in the central module, class RecognitionApp. The core algorithm of the landmark recognition is defined in class EvolvingClassifier. Source code of the class EvolvingClassifier in C++ is attached in the Appendix. Feature selection is also defined in a separate class FeatureSelector. Please note that, sensory data are simply organized into inputs vector in this experiment, however, when a number of features are available, feature selection will be necessary as part of the system in order to choosing the most informative features for the model. In order to deal with the considerable amount of the mathematical calculations using matrices, we also made a static class as a tool for matrix operations.

5.4.4 Results and Analysis

When the input vector, x_k is read by eClass, the algorithm response by labeling the classes $('A','B','C',\ldots)$. Mean while rule-base is updated by either adding a new landmark, and a new class respective to a new corner, replacing an existing landmark (class) or making no change to the fuzzy rule-base structure. In this way, the EFS-based classifier evolves in real-time for example to the following fuzzy rule-base:

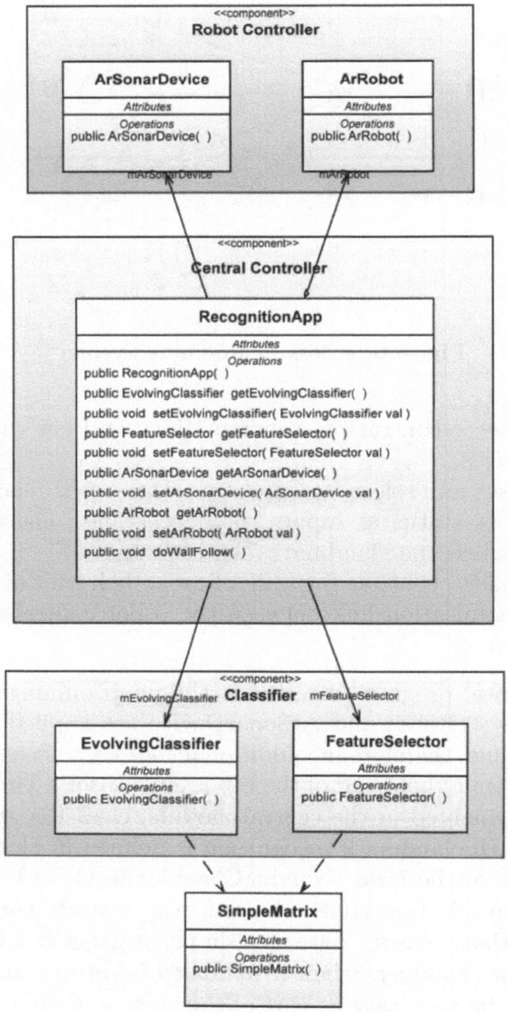

Fig. 5.7. Class diagram of the program

$$R1: IF(\theta\ is\ around\ 0.25)\ AND\ (d\ is\ around\ 0.3000)\ THEN(Corner\ is\ A)$$
$$R2: IF(\theta\ is\ around\ 0.25)\ AND\ (d\ is\ around\ 0.1268)\ THEN(Corner\ is\ B)$$
$$R3: IF(\theta\ is\ around\ 0.75)\ AND\ (d\ is\ around\ 0.0648)\ THEN(Corner\ is\ C)$$
$$R4: IF(\theta\ is\ around\ 0.25)\ AND\ (d\ is\ around\ 0.2357)\ THEN(Corner\ is\ D)$$
$$R5: IF(\theta\ is\ around\ 0.25)\ AND\ (d\ is\ around\ 0.0792)\ THEN(Corner\ is\ E)$$
$$R6: IF(\theta\ is\ around\ 0.75)\ AND\ (d\ is\ around\ 0.1744)\ THEN(Corner\ is\ F)$$
$$R5: IF(\theta\ is\ around\ 0.25)\ AND\ (d\ is\ around\ 0.0371)\ THEN(Corner\ is\ H)$$

(5.21)

After the robot makes one full run in an anti-clockwise direction it was able to recognize successfully 7 out of 8 corners [9] with the remaining corner ('G') incorrectly classified as ('A') due to the close similarity between description of corners 'A' and 'G' (figure 5.8). (Please note that in some rare cases, when the error in sonar reading is big, or when the routine behavior is making extremely big adjustment, the result can be different in the first several runs. This also shows the effects of the sensory precision to the performance.) This result is better comparing to the result reported earlier in [22] where in the similar experiment, 5 out of 8 corners were recognized correctly with 5 features selected as inputs.

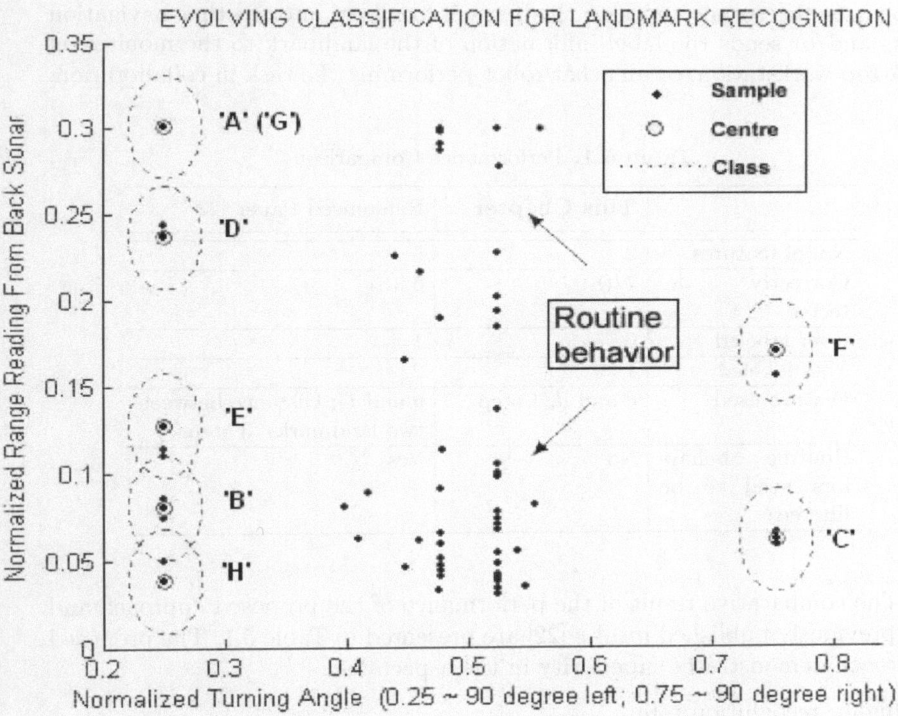

Fig. 5.8. Evolving Classification For Landmark Recognition

The fuzzy rule-based classifier generated in real-time classifies the sensor readings as one of the following 3 cases:

a One of the previously visited landmark (corner) - ClassA- ClassZ;
b New landmark (corner) that has not been visited so far; in this case fuzzy rule-based classifier upgrades its structure automatically;

c Routine behavior - default Class '@';

The corners are the only landmarks available in this simplified experiment. Landmarks are labeled automatically as Classes and the corresponding fuzzy rules practically reflect the corners in the empty office. The landmarks (respectively classes and fuzzy rules) emerges as a result of the real-time experiment based solely on the data. It should be emphasized that the number of landmarks in the rule-base are not pre-defined.

The fuzzy rule base is generated in real-time 'on the fly' and 'from scratch' based on the sensor readings. Seven classes where formed during the first circle around the empty unexplored previously office. They are corresponding to the seven real corners exists in the experimental environment which has eight corners in total. The second time, when the robot goes to a visited corner, the classifier matches the data from the sensors to an existing class, consequently recognizes the corner and uses this for self-localization or further navigation tasks and/or sends the label information of the landmark to the monitoring desk-top workstation, or an other robot performing the task in collaboration.

Table 5.1. Performance Comparison

	This Chapter	Referenced Paper [22]
No. of features	2	5
Correctly detected	7/(8)	5/(8)
Over labeled	0	1
Miss labeled	1	2
Feature used	θ and d, 1 step	θ and D: Distance between two landmarks, 3 steps
Routine behaviors need to be filtered?	No	Yes

The comparative result of the performance of the proposed approach and the previously published results [22] are presented in Table 5.1. The proposed approach demonstrates superiority in the aspects of:

- higher recognition rate;
- higher degree of autonomy;
- higher flexibility (the structure of eClass is not fixed but evolves according to the environment changes)
- higher computational efficiency.

Additional important advantage of the proposed approach is that it has high linguistical interpretability of the information stored in the rule-base which is extracted from the raw sensor readings. In the experiment of [22], a neuro netowrks with 50 neuron is used to support the landmark recognition, which has difficulties in giving explicit and human readable information.

5.5 Further Investigations and Conclusion

In this chapter, a new approach to automatic, fully unsupervised landmark recognition is described. It is based on the evolvable fuzzy rule-based classifier, eClass. A simplified experiment is presented as a case study to illustrate the distinct feature and the application of the real-time classifier. There are some application issues and possible improvements to the algorithm worth further discussing.

5.5.1 Using Different Devices and Selecting Features

In the case study, the environment we chose, an empty office, is simplified compared to the common office environment in normal use. To further simplify the case to give clear demonstration, only two features are chosen as inputs to the model, while more are available, for example, the images from camera, 16 sonar disks of the robot, and laser devices.

In the real world, as the environment is not monotone, applying and properly choosing more available features as inputs can be helpful to increase the precision of the classifier. For example, the color of the the landmarks in the bright environment usually brings distinguishing information. The range information from laser device is more precise comparing to the sonar, and can improve the precision of the result. Using processed images from the camera are also being considered as a potential extension to the experiment, which may enable the proposed application to work in more complicated indoor or outdoor environment. In deep sea environment, the reflection of sonar beam from different objects may bear unique information to identify the objects, and therefore become a good candidate for a feature for the classifier [11].

Specific features have their advantages and disadvantages in different circumstances. Consider in a non-regular shaped enclosure like in figure 5.9, instead of 90°, the 'concave corner' A and 'convex corner' B has very small angles. Consequently, when the robot turning around at these corners, the rotation angle can be very small and very similar to the rotation at routine behaviors. This problem is known in the literature as 'perceptual aliasing' [12]. In this case, using the rotation and back sonar readings as the only inputs (the same as used in the experiment described in the case study,) might not be adequate. In order not to be confused at these two corners, more range information from the sonar, especially readings from the front sonar discs (denoted by $'===' $ in the figure) can be used to differentiate between them.

Please be aware of that, bringing in more features does not always help increasing the performance of the classifier. On one hand, if the feature has no markable impact on differentiate the landmarks, it can be noise information

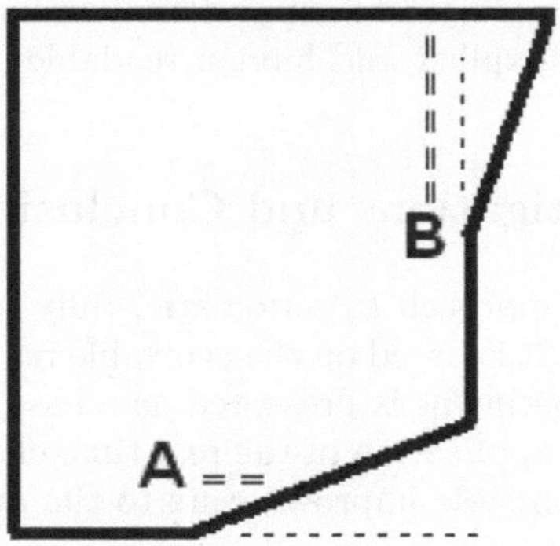

Fig. 5.9. Non-regular shaped enclosure

and therefore, more classes are generated incorrectly. On the other hand, if the feature brought in has large correlation with existing feature, it can weaken the effects of distinct feature in the data space.

Therefore, selecting features is a very important step before inputs are feed to the core part of the evolving fuzzy rule-based classifier. Extensive researches has been carried out in automatical and adaptive selection of features. Further investigation will be directed towards incorporation of such adaptive feature selection into eClass algorithm.

5.5.2 Rules Aggregation

In some cases in the experiment, one corner may be assigned with more than one labels (over-labeling). The reason at the back is that the data space can become over classified and rules with similar antecedents and different consequences can be generated. (See rule 2 and rule 4 in the example rule-base 5.22.) The ability to solve the 'over labeling' (over classification) problem in landmark recognition is important as it gives better performance and increases the interpretability of the rule-base.

$$\begin{aligned}
&R1: IF(\theta\ is\ around\ 0.25)\ AND\ (d\ is\ around\ 0.3000)\ THEN(Corner\ is\ A)\\
&R2: IF(\theta\ is\ around\ 0.25)\ AND\ (d\ is\ around\ 0.1268)\ THEN(Corner\ is\ B)\\
&R3: IF(\theta\ is\ around\ 0.75)\ AND\ (d\ is\ around\ 0.0648)\ THEN(Corner\ is\ C)\\
&R4: IF(\theta\ is\ around\ 0.24)\ AND\ (d\ is\ around\ 0.1270)\ THEN(Corner\ is\ D)
\end{aligned} \tag{5.22}$$

To reduce and further avoid over classification, the rules with very similar antecedents in the fuzzy rule-base need to be properly aggregated into one rule (or at least into a smaller number of rules).

One solution is to aggregate the antecedent in certain way such as taking the mean or choosing one of the two and updating the existing rule. When based on the mean calculation, R2 and R4 are combined in to one rule R2, the updated rule base will be:

$$\begin{aligned} R1 &: IF(\theta\ is\ around\ 0.25)\ AND\ (d\ is\ around\ 0.3000)\ THEN(Corner\ is\ A) \\ R2 &: IF(\theta\ is\ around\ 0.245)\ AND\ (d\ is\ around\ 0.1269)\ THEN(Corner\ is\ B) \\ R3 &: IF(\theta\ is\ around\ 0.75)\ AND\ (d\ is\ around\ 0.0648)\ THEN(Corner\ is\ C) \end{aligned} \tag{5.23}$$

Note that in the case of landmark recognition as the consequent part of the rule is a discrete value, the aggregation on this part of the rule should be prototype based, namely we should choose one of the labels from one of the rules, which comes from a real data point. This is because calculating mean of labels makes no sense. (For example, one can not take mean of corner A and corner B.)

There are other solutions to aggregate the linguistically contradicting rules in the fuzzy rule-base, and intensive research is carried out in this direction. Rule aggregation is even more important in transferring rules among several autonomous robots performing tasks in cooperation.

5.5.3 Applying Variable Radius

In formulae 5.11, the parameter r is a threshold that determines the zone of influence of the class focal point of a class. In reality, it represents the tolerance of a landmark in the environment. The tolerance varies on different landmarks and on different features. For a very distinguishing landmark, this parameter can be bigger, otherwise, it should be small, in order to minimize or eliminate the overlap between classes representing landmarks.

Instead of arbitrarily setting a fixed radius for all landmarks in all dimension, it can be calculated variably in real-time in an adaptive mode based on data from the environment using the local spatial density information. Assuming S denotes the the 'support' (or say 'population') of a class, which means the number of sensory readings that has been classified into this class:

$$S^l \leftarrow S^l + 1 \tag{5.24}$$

where l denotes the index of the winning class.

The spatial variance V of the class at dimension j can be calculated recursively by:

$$V_{jk}^i = \frac{1}{S_k^l} \sum_{l=1}^{S_k^l} \|x_{i*} - x_l\|_j^2 ; \quad V_{j1}^i = 1 \tag{5.25}$$

When a new landmark (rule) is added, $N \leftarrow N{+}1$, its local spatial variance is initialized based on the average of the local spatial variance of the previously existing rules:

$$V_{j1}^{N+1} = \frac{1}{N}\sum_{i=1}^{N} V_{jk}^{i}; j = [1, n] \tag{5.26}$$

Based on the local spatial variance information, we can update the radius of each class adaptively and recursively:

$$r_{jk}^{l} = \rho r_{j(k-1)}^{l} + (1-\rho)\sqrt{V_{jk}^{l}};\quad l = \arg\min_{N}^{k} \left\| x_k - x^{i*} \right\|; \tag{5.27}$$

Note that the learning/compatability rate ρ used here is an leverage set to balance the adaptiveness of the incoming data, witch is between 0 $\sim$ 100%. A higher ρ leads to faster updating of in the radius with less stability; while a smaller ρ gives more stability but less adaptiveness.

5.6 Summary

The methodology for fully automatic and unsupervised landmark recognition by an evolving fuzzy rule-based classifier has been described in this chapter. Extensive experimentation and simulations has been carried out in an office environment with Pioneer 3DX mobile robot by ActivMedia.

The results illustrate the superiority of the proposed evolving technique for simultaneous classifier design and landmark recognition comparing to the previously published results.

A number of further extensions have been discussed and are to be developed. Therese includes but not limited to:

- Adaptive class radius;
- Rules aggregation;
- Working in outdoor and non-regular environment;
- Properly introducing more inputs from different devises, especially image devices (camera).

References

1. P. Angelov, “An Approach for Fuzzy Rule-base Adaptation using On-line Clustering”, Intern. Journal of Approx. Reasoning, Vol. 35, No3, pp. 275-289, March 2004.
2. S. L. Chiu, “Fuzzy model identification based on cluster estimation,” Journal of Intelligent and Fuzzy Syst.vol.2, pp. 267-278, 1994.

3. Yager R.R., D.P. Filev, "Learning of Fuzzy Rules by Mountain Clustering," Proc. of SPIE Conf. on Application of Fuzzy Logic Technology, Boston, MA, USA, pp.246-254, 1993.
4. Angelov, P., D. Filev, "An approach to on-line identification of evolving Takagi-Sugeno models", IEEE Trans. on Systems, Man and Cybernetics, part B, vol.34, No1, pp. 484-498, 2004.
5. Angelov, P. , D. Filev, "Simpl_eTS: A Simplified Method for Learning Evolving Takagi-Sugeno Fuzzy Models", The 2005 IEEE International Conference on Fuzzy Systems FUZZ-IEEE, Reno, Las Vegas, USA, 22-25 May 2005, pp.1068-1073
6. R.O Duda, P.E. Hart and D. G. Stork, Pattern Classification, 2nd edition, John Willey and Sons Inc., New York, USA, 2001
7. Q. Jackson, D.A. Landgrebe, An adaptive classifier design for high-dimensional data analysis with a limited training data set, Geosciences and Remote Sensing, Vol 39 (12), pp. 2664-2679, 2001.
8. L. Bull, T. Kovacs, Foundations of Learning Classifier Systems, Heidelberg, Germany, Springer Verlag, 2005.
9. X. Zhou, P. Angelov, "Real-time joint Landmark Recognition and Classifier Generation by an Evolving Fuzzy System", IEEE World Congress on Computational Intelligenec, Vancouver, Canada, 16-21 July 2006.
10. P. Angelov, X. Zhou, "Evolving Fuzzy System From Data Streams in Real-Time", 2006 International Symposium on Evolving Fuzzy Systems, Ambleside, UK, 7-9 Sept 2006, pp.29-35.
11. D. Carline, P. Angelov, R. Clifford, Agile Collaborative Autonomous Agents for Robust Underwater Classification Scenarios, Underwater Defense Technology, Amsterdam, June 2005.
12. U. Nehmzow, 'Meaning' through Clustering by Self-Organisation of Spatial and Temporal Information, in C. Nehaniv (ed.), Computation for Metaphors, Analogy and Agents, Lecture Notes in Artificial Intelligence 1562, Springer Verlag, 1999, ISBN 3-540-65959-5.
13. Kohonen,'T., Self-Organization and Associative memory. New York, Springer Verlag, 1984.
14. Kasabov N "Evolving fuzzy neural networks for on-line supervised/unsupervised, knowledge-based learning," IEEE Trans. SMC - part B, Cybernetics Vol.31, pp.902-918, 2001.
15. Fritzke B. "Growing cell structures - a self-organizing network for unsupervised and supervised learning," Neural Networks, vol.7 (9) pp. 1441-1460, 1994.
16. Carpenter G. A., S. Grossberg, J. H. Reynolds, "ARTMAP: Su-pervised real-time learning and classification of non-stationary data by a self-organizing neural network.," Neural Networks, vol4, pp. 565-588, 1991.
17. Huang G.-B., P. Saratchandran, N. Sundarajan, "A generalized growing and pruning RBF (GGAP-RBF) neural network for function approximation," IEEE Trans. on Neural Networks, vol.16 (1), 57-67, 2005.
18. Kasabov N., Q. Song "DENFIS: Dynamic Evolving Neural-Fuzzy Inference System and Its Application for Time-Series Prediction," IEEE Trans. on Fuzzy Systems, Vol.10 (2), pp. 144-154, 2002.
19. Plat J., (1991) A resource allocation network for function interpolation, Neural Computation, vol. 3 (2) 213-225.
20. Silva L., F. Gomide, R. Yager, "Participatory Learning in Fuzzy Clustering," The 2005 IEEE International Conference on Fuzzy Systems FUZZ-IEEE, Reno, Las Vegas, USA, 22-25 May 2005.

21. Pioneer 3DX Operations Manual, Activmedia Robotics, LLC Amherst, NH, 2004, chap 5.
22. Nehmzow, U., T. Smithers and J. Hallam, Location Recognition in a Mobile Robot Using Self-Organising Feature Maps, in G.Schmidt (ed.), Information Processing in Autonomous Mobile Robots, Springer Verlag, 1991.

Appendix: C++ Class EvolvingClassifier

```
/* Take inputs, perform evolving classification,return the
class ID the data sample belongs to (subscript of focal point) */
int EvolvingClassifier::DoClassification(vector<double>inputSample)
{
    int classID = -1;
    if(k == 0) //if it is the first inputs, do the initialization
    {
        //——— Stage 1: Initialize focal point set ———//
        k++; //data sample k=1
        R = 0;    //update focal point (rule, class centre) number
        x = inputSample; //take the input
        xPotentialG = 1.0;    //Set potential of first sample to 1

        //init parameters (bk, fk) for recursive calculation of potential
        bk=0;
        vector<double>tmpInit;
        for (int i=0;i<sampleDim; i++)
            tmpInit.push_back(0);
        fk = tmpInit;    //init fk

        classID = 0; //Default class for routine behavior
    }
    else //if not first input, do classification procedure
    {
        k++; //new cycle
        //save old parmeters for recursive calculation
        bk_1 = bk;
        fk_1 = fk;
        x_1 = x;

        //———Stage 2: Take the inputs———//
        x = inputSample;

        //——— Stage 3:Recursive calculation of potential ———//
        /* ————————————————————
        * xPotentialG = 1 - ak + (2*c - bk) /(k - 1)
        * ————————————————————
        * b[k] = b[k-1] + SUM j=1 n xj[k-1]*xj[k-1]
        * fj[k] = fj[k-1] + xj[k-1]
        * c[k] = c[k-1] + fj[k] * xj[k]
        * a[k] = SUM j=1 n xj[k]*xj[k]
        *————————————————————-*/
        //b[k] = b[k-1] + SUM j=1 n xj[k-1]*xj[k-1]
```

```
double tmpSum=0;
for (unsigned int j=0;j<x.size();j++)
    tmpSum = tmpSum + x_1[j] * x_1[j];
bk = bk_1 + tmpSum;

//fj[k] = fj[k-1] + xj[k-1]
for(unsigned int j=0;j<x_1.size(); j++)
fk[j] = fk_1[j] + x_1[j];

//c[k] = c[k-1] + fj[k] * xj[k]
ck=0;
for(unsigned int j=0;j<x.size(); j++)
    ck = ck + fk[j] * x[j];

//a[k] = SUM j=1 n xj[k]*xj[k]
double ak=0;
for(unsigned int j=0;j<x.size(); j++)
    ak = ak + x[j] * x[j];

//potential
xPotentialG = 1 - ak + (2*ck - bk) /(k - 1);
cout <<xPotentialG <<endl; //out put potential for test

//——— Stage 4: Rule base updating ———//
/* ————————————————————————
* condition A: xPotentialG <p_
* condition B: min(sqrt((x-xi*)^2)) <0.5r
*
* A&&B: replace existing nearest focal point
* A&&!B: new focal point (new rule)
* !A&&B: included in an existing class
* ————————————————————————*/
double p_ = 0.935; // threshold for potential

// condition A: potential lower than threshold
bool A = xPotentialG <p_;

// conditoin B: Close enough to an existing class centre
bool B;
double minDelta = (double) N_A;
int nearestRule = -1; //initialy not exists
int disSpaceDim = x.size();

if(R==0)// If no existing rule
{
```

```
        minDelta = 9999999;//set as infinitive big value
        nearestRule = 88888;//set as infinitive big value
    }
    else // find nearest focal point (rule) and the distance
        for (int i=0; i<R; i++)
        {
            double distance = 0; //sqrt((x-xFocal)^2))
            for(int j=0; j<disSpaceDim; j++)
                distance = distance + (x[j]-xFocal.getAt(i,j)) *
                                (x[j]-xFocal.getAt(i,j));
            distance = sqrt(distance);
            if (distance <minDelta)
            {
                minDelta = distance;
                nearestRule =i;
            }
        }
    B = (minDelta <r/2); //close enough to the centre of the nearest focal
point

    //Rule base updating
    if(A && B)
    {//replace nearest focal points
        if(R>1)
        {
            xFocal.replaceRow(nearestRule,x);
            cout <<"REPLACED:" <<(nearestRule+1) <<"sample No."
<<k
                <<"Total Rule:" <<R <<endl ;
        }
        classID = nearestRule+1;
    }
    else if(A)
    {//add new rule
        xFocal.appendRow(x);
        R=R+1;
        cout <<"ADD NEW:" <<R <<"sample No." <<k <<"Total
Rule:"
            <<R <<endl;
        classID = R;
    }
    else if(minDelta <r) //within the radius of a class
    {//assign to an existing class
        cout <<"INCLUDED:" <<(nearestRule+1) <<"sample No." <<k
            <<"Total Rule:" <<R <<endl;
```

```
            classID = nearestRule+1;
        }
    }//else if k == 0
    return classID;
}
```

Part II

Learning Mobile Robots

6

Reinforcement Learning for Autonomous Robotic Fish

Jindong Liu, Lynne E. Parker, and Raj Madhavan

Department of Computer Science, University of Essex,
Wivenhoe Park, Colchester, United Kingdom,
(jliua, dgu, hhu)@essex.ac.uk

The chapter discusses applications of reinforcement learning in an autonomous robotic fish, called Aifi. A three-layer architecture is developed to control it. The bottom layer consists of several primary swim patterns. A sample-based policy gradient learning algorithm is used in this bottom layer to evolve swim patterns. The middle layer consists of a group of behaviours which are designed for specific tasks. The top layer is a Markov Decision Process (MDP), which is used for the planning purpose. The behaviour coordination is conducted by building a MDP in the top layer. A state-based reinforcement learning algorithm, Q-learning in particular, is applied in the top layer to find an optimal planning policy for a specific task. Both simulated and real experiments show good feasibility and performance of the proposed learning algorithms.

6.1 Introduction

The Human Centered Robotics Research Group at Essex has developed a number of robotic fishes since April 2003. Different from other robotic fish projects, we focus on realizing multiform fish-like behaviours and machine intelligence on our robotic fish. The aim of our research project is to make the robotic fish, named Aifi, "grow" from "baby", which is able to learn the best control parameters for its variant behaviours and learn to adapt itself to changes in its environment, such as variable water current and moving obstacles.

To achieve goal-oriented tasks and fast response ability to the dynamics in environments, Aifi is controlled based on a three-layer hybrid architecture (see Figure 6.1). From bottom to top, it comprises a *swim pattern* layer, a *behaviour* layer and a *cognitive* layer. The *swim pattern* layer classifies the swimming motion of robotic fish into several basic swimming elements, called *Swim Patterns*, which interpret the commands from the *behaviour layer* into the low

J. Liu et al.: *Reinforcement Learning for Autonomous Robotic Fish*, Studies in Computational Intelligence (SCI) **50**, 121–135 (2007)
www.springerlink.com

level motion control. It consists of *cruise-in-straight*, *sharp-turning*, *cruise-in-turning* and *ascent/descent*. The *behaviour* layer is designed to quickly response to the sensor data and direct Aifi to apply one of swim patterns. It includes several individual behaviours: *obstacle-avoiding*, *wall-following*, *goal-seeking*, *keep-level*, *wandering*, etc. The *cognitive* layer extracts robotic fish status from the sensor data and conducts task-oriented reasoning and planning. In the *cognitive* layer, it changes the coordination parameters which are used to coordinate all individual behaviours in the *behaviour* layer.

Within this layered architecture, machine learning can be conducted separately at each layer. In the *swim pattern* layer, each swim pattern actually is represented by a series of kinematic functions of motors, which are embedded in the tail joints. The learning is designed to adjust the parameters of kinematic functions to achieve improved performance. In the *behaviour* layer, behaviours are optimized according to the encoding method of behaviours. For example, if behaviours are encoded by fuzzy logic controllers, the learning algorithm will be applied to fuzzy rules and fuzzy function parameters. In the *cognitive* layer, learning algorithms are designed to update the parameters of reasoning and planning. In this chapter, we consider the learning algorithms for the *swim pattern* layer and the *cognitive* layer.

Due to the immature fish swimming mechanism and the variety of robotic fish mechanical structures, it is much difficult to build proper models for swim patterns. As a result, most of the control parameters of robotic fish swim patterns are tuned manually, which rely on the good human expertise. This manual tuning process is normally time-consuming if an optimum solution or sub-optimum solution in practice is concerned. Additionally, the parameters tuned in such a way can only be adapted to a static environment and could not perform well if the environment changes.

Alternatively, various model-free machine learning techniques have been adopted in many bio-mimetic robot projects to find the optimized control parameters, such as biped robots [11] and Sony Aibo robots [4]. Purely policy gradient reinforcement learning was originally proposed in REINFORCE [10] where policy gradient descent was used to update policy parameters. It was further extended to include value iteration in [2] by defining errors as payoffs. In [8], authors proposed a reinforcement learning algorithm without estimating a value function. In these implementations, the policy updating is converted to the parameter updating by making the policy parameterized by a set of parameters. As it is convenient to use experiment samples to find the gradient of the learned policy with respect to parameters, sample based policy gradient reinforcement learning was successfully applied in achieving fast locomotion for the Sony dog's gaits in [5] and an autonomous robot navigation controller in [3].

In this chapter, we adopt two kinds of *Reinforcement Learning(RL)* algorithms as the basic self-learning methods for Aifi. One is a sample-based policy gradient learning, which is used to optimize the control parameters in the swim pattern layer. Another is a state-based RL, which is used to find a mapping between discrete states and actions in the cognitive layer.

The rest of this chapter is organized as follows. Section 6.2 gives a brief description of Aifi. Section 6.3 presents the implementation of the policy reinforcement learning of swim pattern control parameters. Section 6.4 addresses a typical state-based reinforcement learning in the cognitive layer. In Section 6.5, some simulated experiments and real tests are given to show the feasibility and performance of our method. Summary and future work are given in Section 6.6.

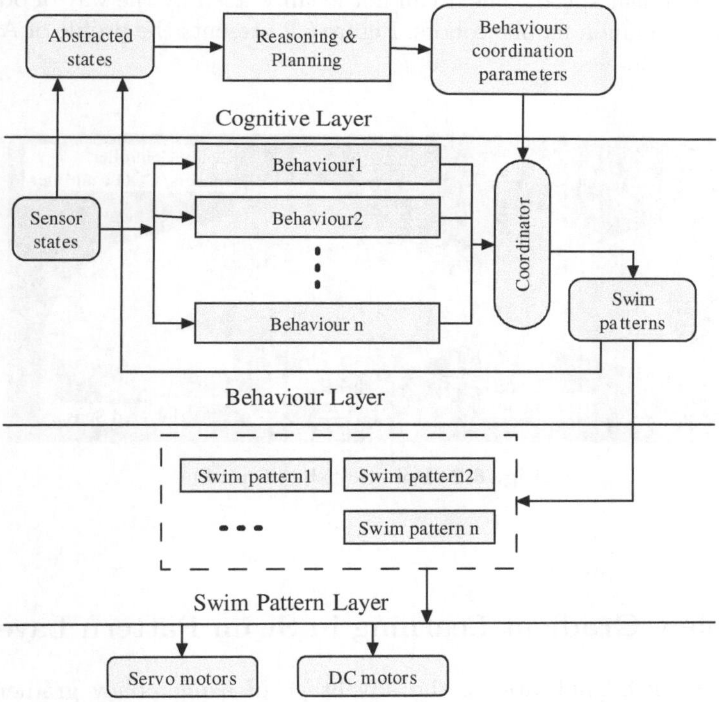

Fig. 6.1. Control structure of Aifi

6.2 Introduction of Robotic Fish-Aifi

Aifi is about 50 cm in length, 20 cm in height and 12 cm in width. It has three joints in its tail which is controlled by three servo motors. Additionally, one DC motor controls its center of gravity position and one mini-pump manages its buoyancy. The center processing unit is a cutting-edge micro-computer, Gumstix [1] which is responsible for all autonomous control computations. Aifi is equipped with several kinds of sensors to response to the dynamical changes in its environment, its position in the tank, the robot attitude and the internal status (e.g. the battery voltage). A standard configuration of Aifi includes four infrared sensors, one dual-axis accelerometer/inclinometer, one piezoelectric vibrating gyroscope, one water pressure sensor, three electric current sensors and three servo turning angle sensors. It is able to sense obstacles around it within a range of 40cm and its depth in the tank. It also can perceive the pitch/roll angle, the one-order derivative of the yaw angle, the turning angle of the tail joints and the power consumption on them. However, Aifi has no ability to localize itself in the horizontal plane because it has no sensor to measure its linear speed, thus it can not localize itself by the way of odometer used in the common mobile robots. Figure 6.2 presents the profile of Aifi used in this research.

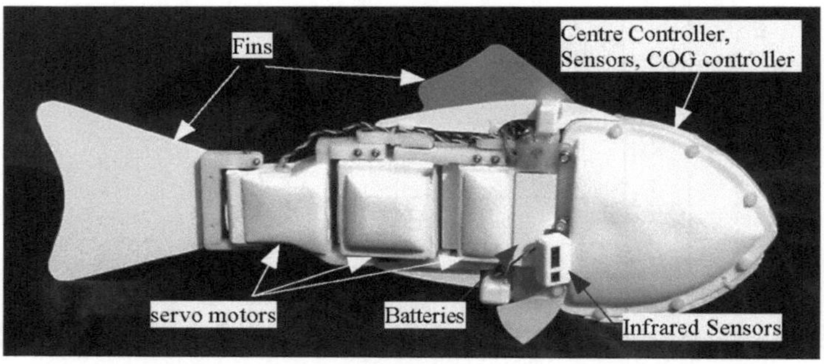

Fig. 6.2. Robotic fish-Aifi profile

6.3 Policy Gradient Learning in Swim Pattern Layer

For robotic fish applications, the advantage of using policy gradient reinforcement learning is that it can integrate the prior knowledge with later autonomous learned experience. The prior knowledge can eliminate the unreasonable parameter selection and limit the learning trace in a narrow feasible parameter space. For example, the largest turning amplitude of each joint is limited by their mechanical design. The maximum or minimum turning speed is both prior decided by the motor type and biological observation on real fishes. All of these knowledge belongs to prior knowledge. The more kinds of prior knowledge we have, the closer the initial value is to optimal. In summary, the prior knowledge is applied to set initial values and the scopes of learning parameters. This integration can shorten the learning time. For instance, the control parameters of a robotic fish can be firstly tuned manually based on any prior experience that is available. After the manual tuning, the fine tuning can be implemented by using a policy gradient reinforcement learning algorithm. Assume that the policy is differentiable with respect to each parameter, the autonomous learning is started from the manual tuning parameters. It estimates the policy's gradients in the parameter space and then updates the parameters to coverage to a local optimum.

First, we define the learning objective for each of robotic fish swim patterns according to their functions. In the policy gradient reinforcement learning, the objective is viewed as the payoff from the environment or the score of the policy function. For example, the maximum turning angle is defined as the objective of *sharp-turning* swim pattern. The payoff indicates how much benefit an agent, i.e. a robotic fish, can receive from its environment after it applies one policy. Normally, the payoff can be measured by sensors that are either on-line or off-line. In our situation, the linear speed is measured by an overhead camera; the angular speed and the power consumption are measured by an embedded compass and an electric current sensor respectively.

A policy π is defined as a probability distribution which is parameterized by the parameters extracted from the control parameters of robotic fish swim patterns. We denote these parameters as $\Theta = \{\theta^1, \ldots, \theta^N\}$. The discounted infinite payoff for this policy is defined as follows:

$$V(\pi) = \sum_{t=0}^{\infty} \gamma^t E[r_t] \tag{6.1}$$

where $\gamma(0 < \gamma < 1)$ is a discount factor, r_t is a payoff and $E[r_t]$ is the expectation of r_t.

Once we obtain an estimate of the discounted infinite payoff gradient with respect to the policy parameters $\frac{\partial V(\pi)}{\partial \theta^i}$, then the policy parameters can be updated by using the following equation:

$$\theta_{t+1}^i = \theta_t^i + \alpha_t \frac{\partial V(\pi)}{\partial \theta^i} \tag{6.2}$$

where α_t is the evolution step.

Due to the lack of the formal expression of the policy π, we can not compute the gradient $\frac{\partial V(\pi)}{\partial \theta^i}$ directly. Instead we use its estimates $\Delta V_{\theta^i}(\pi)$, which can be obtained from samplings on the policy distribution. To avoid large variance occurs in the learning in practical applications, we use the direction of the estimated gradient in the parameter update equation (6.2):

$$\theta_{t+1}^i = \theta_t^i + \alpha_t \frac{\Delta V_{\theta^i}(\pi)}{|\Delta V_{\theta^i}(\pi)|} = \theta_t^i + \alpha_t \eta_t^i \tag{6.3}$$

where $\eta_t^i = \frac{\Delta V_{\theta^i}(\pi)}{|\Delta V_{\theta^i}(\pi)|}$ is the direction of the estimated gradient.

Assume that at episode t the policy is π_t and the parameter vector of π_t is Θ_t. To update Θ_t by Equation (6.3), we introduce terms *direction payoffs* D_t^{+i}, D_t^{0i} and D_t^{-i} to indicate the accumulated payoff in the updating direction "positive", "none" and "negative". They are updated with a discount rate β as follows:

$$D_t^{pi} = \beta D_t^{pi} + (1-\beta) g_t^{pi}, p \in \{+,0,-\} \tag{6.4}$$

where g_t^{pi} is the virtual payoff in the updating direction p of parameter θ^i. It is obtained from sampling by the following process.

Starting from Θ_t, we randomly generate m parameter trials $\{\Theta_t^1, \ldots, \Theta_t^m\}$ around Θ_t by using the perturbation $\Theta_t^j = \Theta_t + \Delta\Theta_t^j$. $\Delta\Theta_t^j$ is defined as follows:

$$\begin{aligned} \Delta\Theta_t^j &= \{0^{j,1}, \ldots, 0^{j,n-1}, \Delta\theta_t^{j,n}, 0^{j,n+1}, \ldots, 0^{j,N}\} \\ &(j = 1...m), n = random(1, N) \end{aligned} \tag{6.5}$$

where superscript j, i denotes the perturbation for ith parameter in jth trial. $random(1, N)$ generates a random integral number between 1 and N in the uniform distribution.

To eliminate the interaction between the perturbations of two parameters, only one parameter is chosen to have the perturbation in each trial. Now, there are m policies close to the initial policy $\pi_t = f(\Theta_t)$: $\{\pi_t^1, \ldots, \pi_t^m\}$. Note that $\Delta\theta_t^{j,n}$ is chosen randomly to be either $+\varepsilon\theta_t^n$ or $-\varepsilon\theta_t^i$. The perturbation step ε is currently fixed for all parameters. Each trial is repeated k times to get the average of payoffs as the expectation value, i.e. $E[r_t^j]$. $E[r_t^j]$ is accumulated together to get the payoff sum S_t^{+i} and S_t^{-i} according to n and $\Delta\theta_t^{j,n}$ as follows:

$$\begin{cases} S_t^{+i} = S_t^{+i} + E[r_t^j], \text{ if } n = i \text{ and } \Delta\theta_t^{j,n} = +\varepsilon\theta_t^n \\ S_t^{-i} = S_t^{-i} + E[r_t^j], \text{ if } n = i \text{ and } \Delta\theta_t^{j,n} = -\varepsilon\theta_t^n \end{cases} \tag{6.6}$$

Then we compute the average payoffs A_t^{+i} and A_t^{-i} for S_t^{+i} and S_t^{-i} respectively. Without any perturbation, we apply the policy π_t by k times and get $A_t^{0i} = E[r_t]$. Now, the g_t^{pi} is calculated by the following rule:

$$g_t^{pi} = \begin{cases} 1 & \text{if} A_t^{pi} = \max\{A_t^{+i}, A_t^{0i}, A_t^{-i}\} \\ 0 & otherwise \\ -1 & \text{if} A_t^{pi} = \min\{A_t^{+i}, A_t^{0i}, A_t^{-i}\} \end{cases}, p = \{+,0,-\} \tag{6.7}$$

And η_t^i is calculated by the following rule:

$$\eta_t^i = \begin{cases} 1 & \text{if} D_t^{+i} = \max\{D_t^{+i}, D_t^{0i}, D_t^{-i}\} \\ 0 & \text{if} D_t^{0i} = \max\{D_t^{+i}, D_t^{0i}, D_t^{-i}\} \\ -1 & \text{if} D_t^{-i} = \max\{D_t^{+i}, D_t^{0i}, D_t^{-i}\} \end{cases} \tag{6.8}$$

Once we get g_t^{pi}, D_t^{pi} is updated by Equation (6.4)

Finally, the parameters are updated by (6.3). The updated parameters construct an updated policy π_{t+1} which is the base point of the learning in the next episode $t+1$. The learning will be terminated when t is larger than episode limitation T_E or the termination condition of Equation (6.9) is satisfied in recent l steps ($l > 4$).

$$|E[r_t] - E[r_{t-1}]| < \tau \tag{6.9}$$

To speed up the learning process, an adaptive rate α_t is adopted here to replace the fixed α_t. It is adjusted according to the changing of $E[r_t]$. Suppose that the termination condition (6.9) is satisfied in the latest h episodes ($h < l$), α_t is adjusted as follows:

$$\alpha_t = \begin{cases} \lambda_1 \alpha_{t-1} \; if\, h = l-2 \\ \lambda_2 \alpha_{t-1} \; if\, h = l-3 \\ \lambda_3 \alpha_{t-1} \; if\, h = l-4 \end{cases} \tag{6.10}$$

where $0 < \lambda_1 < \lambda_2 < \lambda_3 < 1$. They are chosen arbitrarily as 0.7, 0.8, 0.9 representatively.

In this way, a larger learning rate can be used at the beginning and the oscillation at the later stage of the learning could be reduced. Algorithm 6.1 shows the policy gradient reinforcement learning algorithm that we have designed for the parameter updating of robotic fish swim patterns.

Algorithm 6.1 The policy gradient reinforcement learning algorithm

Initialize: $\Theta = \Theta_0$, $D_t^{pi} = 0$
while $t <= T_E$ **do**
  generate m trial policies π_t^j by perturbation (6.5) and reset S_t^{+i}, S_t^{-i} to 0;
  repeat π_t^j on Aifi for k times, get $E[r_t^j]$;
  classify $E[r_t^j]$, get payoff sum S_t^{+i}, S_t^{-i} and average payoff A_t^{+j}, A_t^{-j} by (6.6);
  get A_t^{0i} by make trial of π_t without perturbation;
  calculate g_t^{pi} and η_t^i by (6.7) and (6.8);
  $\theta_{t+1}^i \leftarrow \theta_t^i + \alpha_t \eta_t^i$
  get h, where the condition (6.9) is satisfied in latest h episodes;
  if $h < l$ **then** update α_t by (6.10)
  else terminate the learning process.
  endif
  $t = t + 1$;
endwhile (end of one episode)

6.4 State-based Reinforcement Learning in Cognitive Layer

In the cognitive layer, behaviours should be coordinated to achieve specific tasks, i.e. the fish should reason or plan its actions according to its current states. A typical RL based planner can be described by three parts: *Action Space*- a set of possible actions, *State Space*- the discrete possible situations of a robot on the way from its initial place to a goal, and a mapping from the state space to the action space. A *Markov Decision Process(MDP) Model* can be used to formally model such a planner. The RL can be used to learn the mapping function in this model.

6.4.1 Action Space and State Space

The cognitive layer aims at organizing activities to accomplish a task. The task could be turning on/off a software switch to the execution of a behaviour or just setting a configuration parameter of a behaviour. Actions in the cognitive layer are denoted as ca. To simplify the complexity, the action space is divided into two independent subspaces: level-plane actions and depth-control actions. The level-plane actions (la_i) is related with all the behaviours which affect the movement in all 2D planes parallelling water surface, for example *follow-wall* behaviour, while depth-control actions va_i can change the parameters of the behaviours which control the swimming depth, for example *keep-level* behaviour. Note that, *avoid-obstacle* behaviour is not listed as one of actions in the cognitive layer because it is a low-level behaviour in the behaviour layer.

The states in the cognitive layer are extracted from the sensor readings. They also include the information about which swim patterns is previously executed. The states come from the sensor readings but don't represent the quantity of individual sensors. They are the high-level condition or mode of Aifi. They reflect the significant physical events which are sensed or recognised by a temporal and spatial combination of several sensors. For example, if the down-facing infrared sensor outputs a higher value and the pressure sensor is larger than a threshold value, they indicate that the fish is near to the bottom of the tank. A set of these kinds of events constitutes the state space. In addition, an event can also be a previous swim pattern.

Formally, each event is denoted by a binary variant bv. Once an event occurs, the related bv is set to 1, otherwise it is clear to 0. Grouping n events $(bv_1, ..., bv_n)$ in an order generates a state cs for the cognitive layer, i.e. $cs \leftarrow bv_1, ..., bv_n$. The state space consists of a 2^n combination of n events. To decrease the size of the state space, these events are divided into two independent subspaces: level-states cs_l which only have relationship with the level actions, and vertical-states cs_v which are connected to the depth control actions.

6.4.2 Markov Decision Process Model

The RL based planner in the cognitive layer can be described by two finite MDPs: a level-MDP Γ_l and a vertical-MDP Γ_v. The former is a model for level-states and level-actions while the latter is for vertical-states and depth-control actions. Assume that Γ_l is defined by level-states cs_l, level-actions la and the one-step dynamics of the environment. The state transitions are described by transition probabilities:

$$P_l = Pr\{cs_{l(t+1)}|cs_{l(t)}, la_{(t)}\} \tag{6.11}$$

The expected value of the next reward given current state and action, $cs_{l(t)}$ and $la_{(t)}$, together with next state, $cs_{l(t+1)}$ is expressed as:

$$R_l = E\{r_{t+1}|cs_{l(t)}, la_{(t)}, cs_{l(t+1)}\} \tag{6.12}$$

A policy π is a mapping from states to actions. The optimal policy π^* maximizes the probability of reaching the goal. The *value* of a state s under a policy π, denoted $V^\pi(s)$, is the expected return when starting in s and following π thereafter. Function V^π is called *state-value function* for policy π. So, for Γ_l, we define $V_l^\pi(cs_l)$ as:

$$V_l^\pi(cs_l) = E_\pi\{R_t|cs_l\} = E_\pi\{\sum_{k=0}^{\infty}\gamma^k r_{t+k+1}\,|cs_l\} \tag{6.13}$$

where $E_\pi\{\}$ represents the expected value given that Aifi follows policy π. At the same time, the value of taking level action la in level-state cs_l is defined under policy π as $Q_l^\pi(cs_l, la)$:

$$Q_l^\pi(cs_l, la) = E_\pi\{R_t|cs_l, la\} = E_\pi\{\sum_{k=0}^{\infty}\gamma^k r_{t+k+1}\,|cs_l, la\} \tag{6.14}$$

For Γ_v, there are similar definitions of P_v, R_v, $V_v^\pi(cs_v)$ and $Q_v^\pi(cs_v, va)$. Given P_l and R_l for all level states cs_l and level actions la, a full description of Γ_l can be obtained. The optimal policy π^* can be found analytically by using *Dynamic Programming* which recursively calculates V_l^π and Q_l^π. If P_l is unknown, modelling techniques can be used to find it by the model-based RL. Alternatively, π^* can be found directly based on R_l through the model-free RL, such as *Monte Carlo method* or *Temporal-Difference Learning (TD Learning)*. In this chapter, *Q-learning* is designed to learn the mapping in the cognitive layer.

The *one-step Q-learning* updated function is as follows [9]:

$$Q(s, a) = Q(s, a) + \alpha\left[r + \gamma \max_{a'} Q(s', a') - Q(s, a)\right] \tag{6.15}$$

where Q represents Q_l^π, s denotes cs_l and a denotes la. α and γ are learning ratios. r is the observed reward when taking action a. s' is the succeeded state after taking action a. ϵ-greedy method is used to generate a from s using the policy derived from Q. A brief process is shown in Algorithm 6.2.

6.5 Experimental Results

6.5.1 Policy Gradient Learning for *Sharp-Turning* Swim Pattern

To prove the feasibility of the policy gradient learning algorithm, Aifi is used to learn the control parameters of the maximum turning angle of *sharp-turning*

Algorithm 6.2 The Q-learning algorithm

1. **Initialize:** $Q(s_0, a_0)$ arbitrarily;
2. **While** $t < T_E$ **do**:
3. choose a_t from s_t using the policy derived from Q via ϵ-greedy;
4. Take action a_t, observe new state s_{t+1} and reward r_{t+1};
5. $Q(s_t, a_t) = Q(s_t, a_t) + \alpha \left[r_t + 1 + \gamma \max_a Q(s_{t+1}, a) - Q(s_t, a_t) \right]$;
6. t=t+1;
7. **endwhile**

swim pattern. The *sharp-turning* swim pattern was designed in [7] where 8 key parameters $(\theta_1, \ldots, \theta_8)$ are extracted to mimic the sharp turning of real fish. Although this pattern has a kinematic function and a proximate dynamic function, it is quite difficult to obtain the analytical expression of the turning angle according to θ_i. So the relationship between the turning angle and θ_i is viewed as a blackbox which is described by a policy π with parameters θ_i. The objective of the policy gradient learning here is to find a local optimized policy π^* with which Aifi could be expected to have a largest turning angle by executing the *sharp-turning* swim pattern.

First, we adjusted θ_i manually by prior knowledge. The manual tuned θ_i are set as the initial value Θ_0 of the policy gradient learning algorithm, i.e. $\Theta_0 = \{\theta_1, \ldots, \theta_8\}$. Then the algorithm is started from Θ_0 and follows the step listed in Algorithm 6.1. The turning angle during sharp turning is measured by a compass sensor in the fish head. The instant reward is set to equal the final turning angle once *sharp-turning* finishes. In each episode, $m = 30$ trials are tested and each trial is repeated $k = 3$ times. Figure 6.3 shows the learning result for each episode. Initially the average turning angle is about 50 degrees. After about 90 hours learning, Aifi tried 4500 hard turnings, i.e. at the end of the 50th episode, the turning angle increases to 110 degree. Figure 6.4 shows a video sequence of *sharp-turning* after learning.

6.5.2 Q-learning for *Tank Border Exploration* Task

To test the feasibility of the proposed state based learning in the cognitive layer, a tank border exploration task is implemented by Aifi. The objective is to make Aifi to autonomously explore the tank border. It should follow tank walls in an appointed distance, be able to avoid the corner and other fishes, and keep itself in a desired depth level. Additionally it must keep the wall at its right side. Aifi is supposed to know nothing about its environment. It will learn to explore the tank border by the state based learning from scratch. The action space is structured according to the objective of the task. Four behaviours are chosen and customized from the generic behaviours layer for the level plane action subspace.

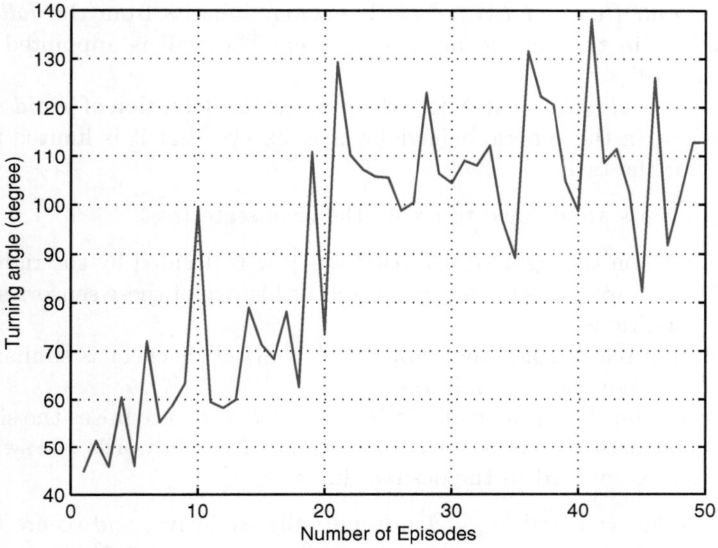

Fig. 6.3. The turning angle of *sharp-turning* swim pattern during learning.

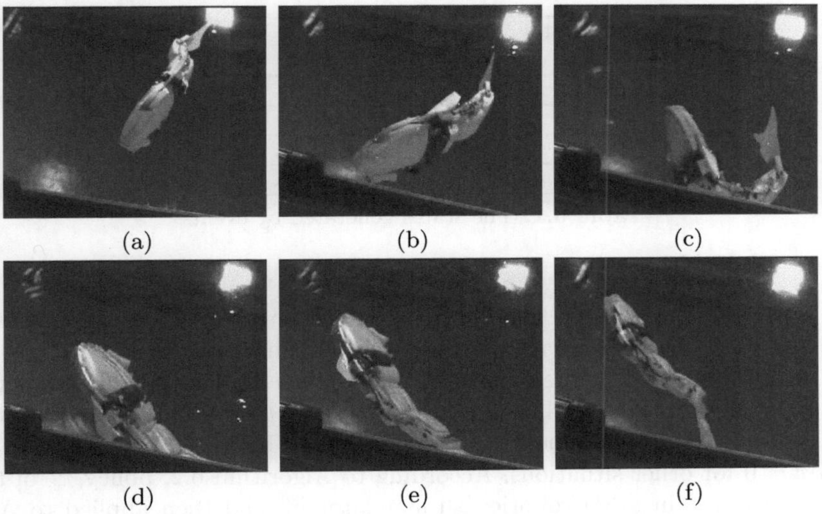

Fig. 6.4. A sharp-turning sequence of Aifi after learning

- *Wander* ($la_1 = WD$): This behaviour is limited on a 2D plane. Aifi randomly selects one of swim patterns from *cruise-straight* and *cruise-in-left/right-turning* to execute.

- *Follow-wall* ($la_2 = FLW$): This behaviour inherits from the *follow-wall* behaviour in the generic behaviour layer. The wall is appointed on the right of Aifi
- *Avoid-obstacle* ($la_3 = AO$): It is the same as the definition of *avoid-obstacle* behaviour in the generic behaviour layer except that it is limited in a 2D plane for the task.

Three events are defined to create the level states(cs_l):

- Is the wall on the right side of Aifi? (bv_1): It is decided by the right, front and left infrared sensors. $bv_1 = 1$ if recent history of these sensors satisfied some conditions.
- Is Aifi in a reasonable range from wall? (bv_2): This event is similar to bv_1 but it has more strict conditions.
- Is the wall on the left side of Aifi (bv_3): Like bv_1, it recognises the situation that the nearest wall is on the left side of fish. In other words, the swimming direction is reversed to the desired direction.

Because $bv_1 = 1$ and $bv_3 = 1$ are mutually exclusive, and so are $bv_1 = 0$ and $bv_2 = 1$, the total number of states is decreased from $8(2^3)$ to 4 as shown in Table 6.1

state cs_l	bv_1	bv_2	bv_3
0	0	N/A	0
1	0	N/A	1
2	1	0	N/A
3	1	1	N/A

Table 6.1. The states generated by events

There is no vertical action subspace in the cognitive layer for this task because the *keep-level* behaviour is capable enough for the task.

In this task, $r = 1$ when the fish in the *follow-wall* behaviour keeps following the wall, $r = -1$ when the fish in the *follow-wall* behaviour loses the wall to follow, $r = -3$ when there is a bumping between the wall and Aifi and $r = 0$ for other situations. According to Algorithm 6.2, policy π^* of the task is learned in a 3D robotic fish simulator [6] and then applied to Aifi. Figure 6.5 presents fish swimming trajectories during learning. It is clearly shown that Aifi is able to keep itself in a proper distance away from the wall after learning. Figure 6.6 gives the history of root mean square (RMS) errors between the learning trajectory and the desired path over 300 trips. The RMS error decreases and converges to a low value as the number of learning trips increases.

Figure 6.7 shows the trajectory of Aifi operated in the real tank, which is recorded from an overhead camera. The arrows in the figure show the heading directions and the circle points indicate the position of Aifi. The time step between two record points is 3 seconds. After Aifi is put into water from the point S with heading direction B_0, it selects the *find-wall* behaviour and executes it until reaching the left wall. Then the *follow-wall* behaviour is triggered. During its swimming, Aifi encounters an obstacle around point B_1. It implements a *sharp-left-turning* swim pattern to avoid it. Then it finds the wall again by the *cruise-in-turning-right* swim pattern. Finally, Aifi spent 105 seconds to swim around the tank one circle and finally reached the point E.

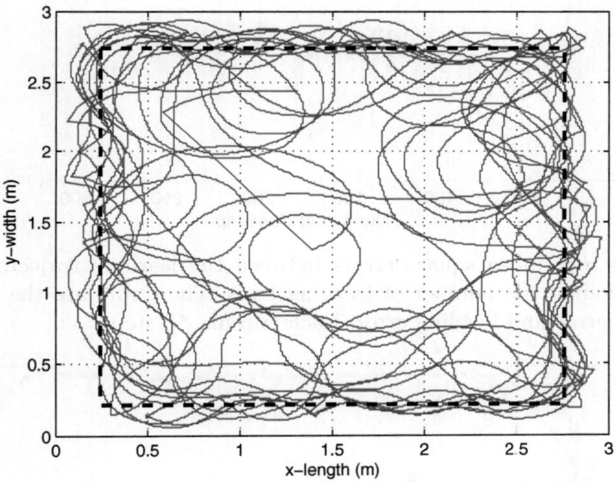

Fig. 6.5. The fish trajectory during learning. Note that the dashed line is the desired path, $\alpha = 0.5$ $\gamma = 0.3$, $\epsilon = 0.01$

6.6 Summary

In this chapter, reinforcement learning is used as learning methods in a layered control architecture of our robotic fish, Aifi. The swim pattern is learned by a sample-based policy gradient learning algorithm in the *swim pattern* layer. The task planning is learned by a state-based RL learning algorithm in the *cognitive* layer. The experimental tests show good performance of both algorithms. In the next step, we will apply reinforcement learning to learning of behaviours in the *behaviour* layer.

References

1. http://www.gumstix.org.

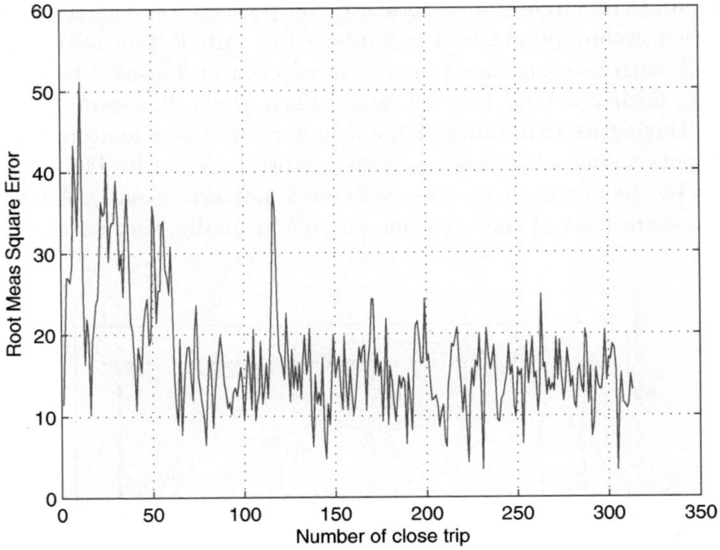

Fig. 6.6. The root mean square errors between the learning trajectory and the desired path against the number of learning trips. The errors are the sum of two parts: position error and heading error. Each trip has 50 steps.

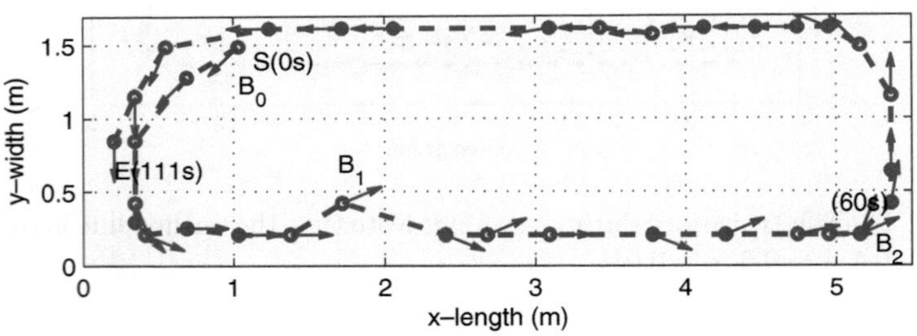

Fig. 6.7. A circular trajectory in the level plane

2. L. C. Barid and A. W. Moore. Gradient descent for general reinforcement learning. In *Proceedings of the International Conference on Advances in neural information processing systems II*, pages 968–974. MIT Press, 1999.
3. G. Z. Grudic, V. Kumar, and L. Ungar. Using policy gradient reinfrocement learning on autonmous robot controllers. In *Proceedings of IEEE/RSJ International Conference on Intelligent Robots and Systems*, pages 406–411, Las Vagas, Navada, USA, Oct 2003.
4. G. Hornby, S. Takamura, J. Yokono, O. Hanagata, T. Yamamoto, and M. Fujita. Evolving robust gaits with AIBO. In *Proceedings of IEEE International Conference on Robotics and Automation*, pages 3040–3045, 2000.
5. N. Kohl and P. Stone. Policy gradient reinforcement learning for fast quadrupedal locomotion. In *Proceedings of IEEE International Conference on Robotics and Automation*, volume 3, pages 2619–2624, May 2004.

6. J. Liu and H. Hu. Building a 3d simulator for autonomous navigation of robotic fishes. In *Proceedings of IEEE/RSJ International Conference on Intelligent Robots and Systems*, pages 613–618, Sendai, Japan, Oct 2004.
7. J. Liu and H. Hu. Mimicry of sharp turning behaviours in a robotic fish. In *Proceedings of IEEE International Conference on Robotics and Automation*, pages 3329–3334, Barcelona, Spain, April 2005.
8. L. Peshkin, K. Kim, N. Meuleau, and L. Kaelbling. Learning to cooperate via policy search. In *Proceedings of the 6th International Conference on Uncertainty in Artificial Intelligence*, pages 307–314, 2000.
9. C. J. C. H. Watkins. *Learning from Delayed Rewards.* PhD thesis, Cambridge University, 1989.
10. R. J. William. Simple statistical gradient-following algorithms for connectionist reinforcement learning. *Machine learning*, 8:229–256, 1992.
11. R. Zhang and P. Vadakkepat. An evolutionary algorithm for trajectory based gait generation of biped robot. In *Proceedings of the International Conference on Computational Intelligence, Robotics and Autonomous Systems*, Singapore, 2003.

7

Module-based Autonomous Learning for Mobile Robots

Esther L. Colombini and Carlos H. C. Ribeiro

NCROMA Research Group, Computer Science Division
Technological Institute of Aeronautics
Praça Marechal Eduardo Gomes, 50 - Vila das Acácias
São José dos Campos, Brazil
esther,carlos@comp.ita.br, http://www.comp.ita.br/ ncroma/

The information available to robots in real tasks is widely distributed both in time and space, requiring the agent to search for relevant information. In this work, we implement a solution that uses qualitative and quantitative knowledge to make robot tasks able to be treated by Reinforcement Learning (RL) algorithms. The steps of this procedure include: 1) to decompose the overall task into smaller ones, using abstractions and macro-operators, thus achieving a discrete action space; 2) to apply a state model representation to achieve both time and state space discretisation; 3) to use quantitative knowledge to design controllers that are able to solve the subtasks; 4) to learn the coordination of these behaviours using RL, more specifically Q-learning. The proposed method was verified on a set of robot tasks using a Khepera robot simulator. Two approaches for state space discretisation were tested, one based on features — that are observation functions of the environment — and the other on states. The learned policies over these two models were compared to a predefined hand-crafted policy. It was found that the learned policy over the state-based discretisation leads quickly to good results, although it can not be applied to complex tasks, where the state space representation becomes computationally unfeasible and a generalisation method has to be applied. The generalisation approach chosen implements the CMAC (Cerebellar Model Articulation Controller) method over the state-based model. The results show that the resulting compact representation allows the learning method to be applied over the state-based model, although the learned policy over the feature-based representation has a better performance.

E. L. Colombini and C. H. C. Ribeiro: *Module-based Autonomous Learning for Mobile Robots*, Studies in Computational Intelligence (SCI) **50**, 137–159 (2007)
www.springerlink.com

7.1 Introduction

Autonomous robots are machines that are built to operate in changing and partially unknown environments. Hence, they can not be programmed to execute predefined action sequences because it is not possible to know in advance what will be the universe of necessary sensorial and motor transformations required by the various situations the robot might encounter.

Reinforcement Learning (RL) is a class of learning suitable for robots when online learning without sufficient prior information about the environment is required. Most RL techniques uses the theory of Markovian Decision Processes (MDP) as their mathematical model, which requires finite state and action spaces. Among the difficulties posed by real robot tasks that have to be overcome to allow them to be treated by RL are: continuous time, continuous action and state spaces and partial observability of states.

As the task to be executed by the robot becomes more complex, it is usually necessary to introduce some form of hierarchy of behaviours which can simply consist in the decomposition of the task into a set of simpler ones. In fact, in recent years, an approach to Artificial Intelligence has been developed which is based on building behaviour-based programs to control situated and embodied robots in changing environments [1]. The design of architectures composed of very simple skills is not easy, nor is the learning of its sequence, as producing an adequate combination of these behaviours is not straight- forward. Furthermore, the controller decomposition introduces the need for determining when to trigger control, i.e. when to re-evaluate the previously selected behaviour and choose a new one.

In this work, qualitative knowledge is used to achieve discretisation over the action space by using abstraction and macro-operators. To acomplish time and state space discretisation two approaches for space representation (state-based and feature-based) are proposed. The first approach is a direct sensorial representation where the other uses observation functions of the environment.

Quantitative knowledge is used to design the controllers that solve the subtasks while RL, more specifically Q-learning, is responsible for learning the coordination of these behaviours. The switching action policies learned over the two approaches for space discretisation are compared to a predefined hand-crafted action policy.

For the cases where the most complex defined tasks could not be executed over the state-based model, a generalisation method called CMAC (Cerebellar Model Articulation Controller) was implemented to reduce the storage and computational requirements for the Q-learning algorithm.

7.1.1 Bibliography Review

Reinforcement Learning [2, 3] is learning by doing. In this approach, the agent learns how to map states to actions by trial and error, without an external supervisor. Since [2] and [4] proposed Markovian Decision Processes (MDPS) to be used as the model for RL analysis a mathematically well-founded theory has been developed for this class of algorithms and many solutions have been proposed [5, 6, 7].

The main problem is that this theoretical basis assumes a finite set of actions and states and a discrete time model where the states should be available for measurement, whereas real robot tasks have infinite state and action spaces, continuous time and due to sensorial limitation are not always measurable. For such situations, for these cases, no complete theoretical solution can be found. In fact, [8] shows that these problems are intractable due to their partial observability.

The use of abstractions, subgoals and macro-operators [9] to decompose tasks into smaller ones, hence allowing large but observable problems to be handled, was applied in planning domains by [10, 11], and turned out to be a very promising approach. [12] showed that if this decomposition is done in a hierarchical manner, it can reduce the problem complexity from exponential to linear.

Macro-operators and learning in planning domains have been approached and related in many contexts, such as: plan acceleration; control modules learning [11, 13], also called learning of macro-operators, and learning of the switching of the particular controllers [14, 15]. [16] studied a more difficult approach: inventing macro-operators, subgoals and hierarchy. Switching control, under the name of hybrid control [17, 18], has also received some attention.

Learning a switching function able to trigger an specific controller at certain time step is the main aspect of this work. In contrast to ours, [13] work has fixed the switching policy and has left to the learning agent the task of finding good controllers. [19, 20, 21] research is based on building behaviour-based programs to robot control where a decision making procedure defines the execution sequence of the predefined modules.

The module concept [22] (operating conditions together with controllers) is well fitted to [16] concept of skill, which has to be learned by the algorithm to help turning the set of tasks complete.

To deal with large or infinite state spaces, the concept of features in RL [23] became one of the tracks that influenced the work of [22]. Although some optimisation (state complexity reduction) is achieved, it can not be expected that working on the features vector will remove partial observability. Issues in learning in partially observable environments have been discussed by [24].

Because of the state explosion encountered on some space discretisations, compact representations are applied to deal with storage, computational and convergence conditions. A natural way to incorporate generalisation into RL methods is based on the use of function approximators, rather than look-up tables, to represent the value function. [13, 25] have demonstrated some control applications were function approximators have succeeded. The great success of [26, 27] on finding good backgammon solution also contributed to the popularity of approximators in RL.

The use of generalisation in Reinforcement Learning is not a new approach. It goes back to the 50's when [28] used RL to adjust the parameters of linear threshold function representing policies. [29] used the theory for learning value functions.

In the 70's, Albus [30] proposed the CMAC approach that was lately described in terms of tile coding by [31]. Tile coding have been applied to some control problems such as [32, 33] as well as reinforcement learning systems [34, 35]. [36] shows a soccer robot application using Q-learning and CMAC.

7.2 Reinforcement Learning

Reinforcement Learning (RL) [2] is a technique that allows an agent to adapt to its environment through the development of an action policy, which determines the action that should be taken in each environmental state in order to maximise (or minimise) a function over a cumulative reinforcement. The reinforcement is a real value that defines the desirability of a state and can be expressed both in terms of rewards or punishments. In RL systems, the a priori domain knowledge incorporated by the designer is minimal and is mostly encapsulated in the reinforcement function.

In RL, convergence is conditional on an infinite number of visits to every possible state of the process the agent must adapt to. The need for simultaneous exploration and best policy execution creates the exploration/exploitation tradeoff, also known as the *dual control problem* [6]. The most common form of treating it is by properly choosing random actions, according to the so-called *Boltzmann exploration strategy* [37]. In this work a simpler approach where the exploration rate decays linearly over time was applied.

A limitation that arises from real robot tasks is the agent inability to completely measure or represent its state. This problem is known as *perceptual aliasing* [38]. When it occurs, different environment situations can be erroneously represented as a single state. There are many methods based on the use of memory or attention to deal with this problem [39].

In RL tasks the rewards define the objective. A poorly chosen reward function can cause the learning system not to converge, or to converge to a policy that does not accomplish the desired task. In [40] it is shown that dense rewards (non-zero values) facilitate clever exploration, which can reduce the search complexity in the course of learning.

7.2.1 Markovian Decision Processes

Nowadays, most of the theory involving Reinforcement Learning is restricted to Markovian Decision Processes (MDPs). To be considered as such, the process has to satisfy the Markov Condition where any observation o made by the agent must be a function only of its last observation and action (plus some random disturbance), i.e. $o_{t+1} = f(o_t, a_t, w_t)$. If this condition is guaranteed, the process can be modeled as a 4-tuple $\langle S, A, P, R\rangle$, where:

- S is a finite set of states
- A is a finite set of actions
- P is the transition probabilities model that maps action-state pairs $P(s_{t+1}|s_t, a_t)$
- R is the reward function $r(s, a)$

Hence, we can estimate the probability to reach new state s' and the reward associated to this move from:

$$P^a_{ss'} = Pr\left\{s_{t+1} = s' | s_t = s, a_t = a\right\} \tag{7.1}$$

$$R^a_{ss'} = E\left\{r_{t+1} | s_t = s, a_t = a, s_{t+1} = s'\right\} \tag{7.2}$$

The objective of learning in this context is to identify an optimal policy. This policy is some function that tells the agent which set of actions should be chosen under certain circumstances. Dynamic Programming methods are the basic algorithms to achieve this optimal policy [7]. They employ a complete world model where the transitions probabilities are known. In this work it will be used a RL variation of these methods that do not need a complete transition probability distribution: the Q-learning algorithm.

7.2.2 Q-learning

Q-learning [31] is the preferred RL algorithm because it provides good experimental results in terms of learning speed and it is a model-free learning for optimal policies.

The iterative process for calculating the optimal policy is done as follows. At time t, the agent:

1. Visits state s_t and selects an action a_t
2. Receives the reinforcement $r_t = r(s_t, a_t)$ and observes the next state s_{t+1}
3. Updates $Q_t(s_t, a_t)$ according to:

$$\begin{aligned} Q_{t+1}(s_t, a_t) &= Q_t(s_t, a_t) \\ &+\alpha_t[r_t + \gamma\hat{V}(s_{t+1}) - Q(s_t, a_t)] \end{aligned} \tag{7.3}$$

4. Repeats steps above until stop criterion is satisfied.

where $\hat{V}(s_{t+1}) = min_a[Q_t(s_{t+1}, a)]$ is the current estimate of the optimal expected cost, α is the learning rate and $\gamma (0 < \gamma < 1)$ is the discount factor that reduces the influence of future expected rewards. Q-learning learns the values of all actions in all states, rather than only representing the policy.

7.2.3 Features

In real robot tasks, the system can not be modelled as a finite MDP as there is no complete state information available for measurements. To deal with this partial observability condition, an extension of MDP is provided which maps the set of states S into a new finite set X, called the observation set. An observation function h is then applied to s, making the whole state s only observable by $h(s)$. Features are the dimensions of $h(s)$, and using observation functions as many-to-one representations produces a way to deal with the problem of state space infiniteness. If extended in a sensible way, the use of features can also deal with partial observability. The feature values depend upon each observation-action pair, and its value does not change as long as the observation-action pair remains unchanged.

It is convenient to work with a set of features, where each one represents an event to be taken into account. The event triggering, i.e. the change of an observation-action pair value leads to time discretisation.

7.2.4 Module-based RL

The idea of using abstraction, subgoals and macro-operators [9] to decompose tasks into smaller ones, hence allowing large but observable problems to be handled. It was applied in planning domains by [10, 11], and turned out to be a very promising approach.

It consists of using qualitative knowledge about the problem to divide it into smaller ones again and again, through the definition of subgoals, sub-subgoals, etc, until reaching a level where these small problems can be represented by simple routines. The entire problem is then solved by the re-combination of the small problems until reaching the main task. This main task, made of smaller ones, exists in planning domains under the name of *macro-actions.* [12] showed that if this decomposition is done in a hierarchical manner, it can reduce the problem complexity from exponential to linear. In the end, we have a set of macro-actions associated with their specific subgoals. A subgoal can also be defined in robotics as a desired behaviour.

As it can be noted in Figure 7.1, the design phase is responsible for manipulating this knowledge by implementing macro-actions as local controllers. These macro-actions should be applied under well defined and measurable conditions.

The controllers - macro-action implementation - together with their operating conditions -basic features set - are here called Modules. Although a well defined set of conditions has been defined for each module, they are not exclusive, i.e. many controllers can be active under the same circumstances. To overcome this scenario, a switching function should be applied. The decision should be taken on the basis of the state of operating conditions plus some possible additional filters that, together, represent the feature vector for the switching mechanism of Figure 7.2 [22]. The last phase showed in

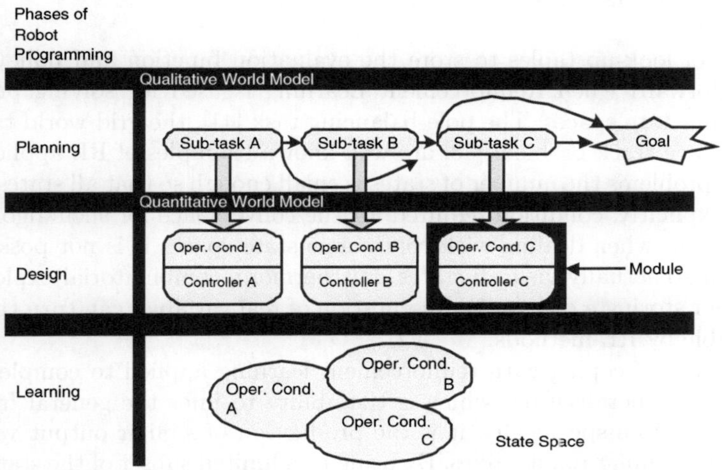

Fig. 7.1. Phases of the used approach (adapted from [22])

Figure 7.1, represents the application of RL as the mechanism to learn the above mentioned switching function.

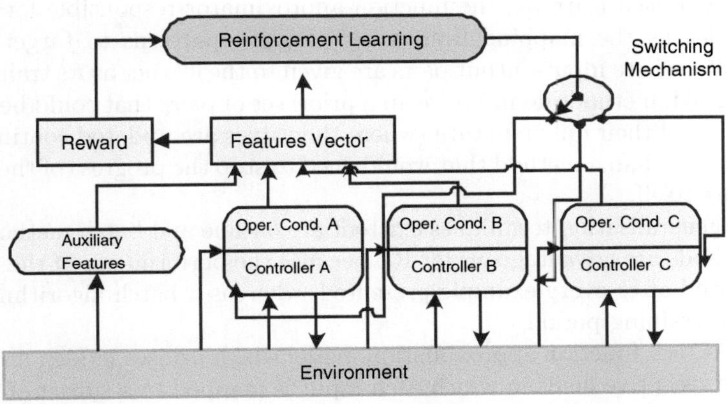

Fig. 7.2. The control and learning mechanism (adapted from [22])

7.3 Generalisation

The use of look-up tables to store the evaluation function and policy seems straightforward when Reinforcement Learning is used for solving problems with finite state spaces. The pole-balancing task [41], the grid world task [42] and the race track problem [43] are well-known examples of RL applications. In these problems the number of states is small enough so that all states can be visited explicitly, condition required for the convergence for such algorithms.

However, when dealing with continuous state spaces it is not possible for the agent to actually visit all states. Furthermore, combinatorial explosion of a table for storing a discrete representation of real variables can turn the tasks intractable by RL methods.

One way of coping with reinforcement learning applied to complex tasks is by using generalisation, which is the ability to infer the general from the particular. More specifically, it is the production of similar output values in response to similar input values. By using it, a limited subset of the state space is useful to generate approximated information about a much larger number of states. This approximated information is generated by a function approximator, that takes examples from a target function mapping and attempts to generalise from it to construct an approximator for the entire function.

Two clear benefits of the application of generalisation in the learning process are that it is speed up and the storage space required is much smaller then by using look-up tables.

In supervised learning, the function approximator responsible for generalisation learns the mapping from a set of input patterns to target output values [35], where input-output pairs are given to the learner as its training set. However, RL methods do not have an a priori set of pairs that could be used in a tracing and their online nature (where the values are updated continuously) require more than a method that would need to stop the progress of the system to update itself.

The generalisation techniques existent are on-line and batch methods. On-line methods are advantageous for RL because the performance of the method is improved after every example presented, whereas a batch algorithm would require a training period.

CMAC is a function approximation model which utilises hidden units with localised receptive fields in which each input is mapped to a subset of weights whose values are summed up to produce the outputs. Because CMAC together with a Q-learning implementation has been treated by many in the literature [31, 34, 44] and its local generalisation ability has been beneficial in RL, it was the approach chosen for implementation in this work.

7.3.1 Cerebellar Model Articulation Controller (CMAC)

In the early 70's, James Albus [30, 45] modelled the human cerebellum function of information processing by using a neural structure named CMAC

(Cerebellar Model Articulation Controller). It is a coarse-coding structure that consists of an associative memory neural network in which each input is mapped to a subset of weights whose values are summed up to produce the outputs.

In CMAC, a set of overlapping, multi-dimensional layers, also called receptive fields or tillings, have a finite size and are defined by quantising functions. Each element of a tilling is called a tile and it represents a receptive field for one binary feature. The set of excited receptive fields of corresponding quantising functions in all dimensions of the input space are combined to define a hypercube in the multi-dimensional input space, that corresponds to one component of the output value. Because of the many quantising functions existent in the input dimension, many hypercubes are affected. The output is then computed by summing the contributions from the components activated.

Figure 7.3 shows a general representation of the CMAC approach that will be detailed in section 7.3.1.

One of the benefits of CMAC is that it has local generalisation, where similar inputs produce similar outputs and far located inputs produce nearly independent outputs. Furthermore, it is an alternative to the backpropagation method used for learning in neural networks because it has faster convergence [35], meaning that the number of iterations required to converge is much smaller in the CMAC approach, thus allowing it to be used in real-time settings.

CMAC working

The mode CMAC operates and its generalisation capability is demonstrated in figure 7.3. In this representation two state variables s_1 and s_2, that compose the state space $s = [s_1, s_2]$, are provided. For each input dimension there are four quantising functions ($k = 4$) with six resolution elements each, also called tiles ($Q_{i,j} = 6, i = 1,2, j = 1,2,3,4$). Actually, the number of resolution elements does not have necessarily to be the same for each quantising function in each dimension.

Consider then two points A and B whose input state values are $s_A = [2.35; 3.36]$ and $s_B = [2.55; 3.56]$. The first point A has been visited and it had its weights updated, where B is a nearby point in the state space that has never been visited.

When the values of the input for point A are quantised by the quantising functions f_{lN}, with $l = \{1,2,3,4\}$ representing the lth quantising function in the input space dimension s_N and $N = 1,2$, the set of resolution elements achieved is represented by: $m^*_{A1} = \{D, J, P, V\}$ and $m^*_{A2} = \{c, j, o, u\}$, where the letters represent the selected resolution element.

Then, the letters for resolution elements from corresponding quantising functions are concatenated to form the set of active hypercubes $A^*_A = \{\{Dc\}, \{Jj\}, \{Po\}, \{Vu\}\}$, where the output y_A is the sum of the weights indexed by these components $y_A = w[\{Dc\}] + w[\{Jj\}] + w[\{Po\}] + w[\{Vu\}]$.

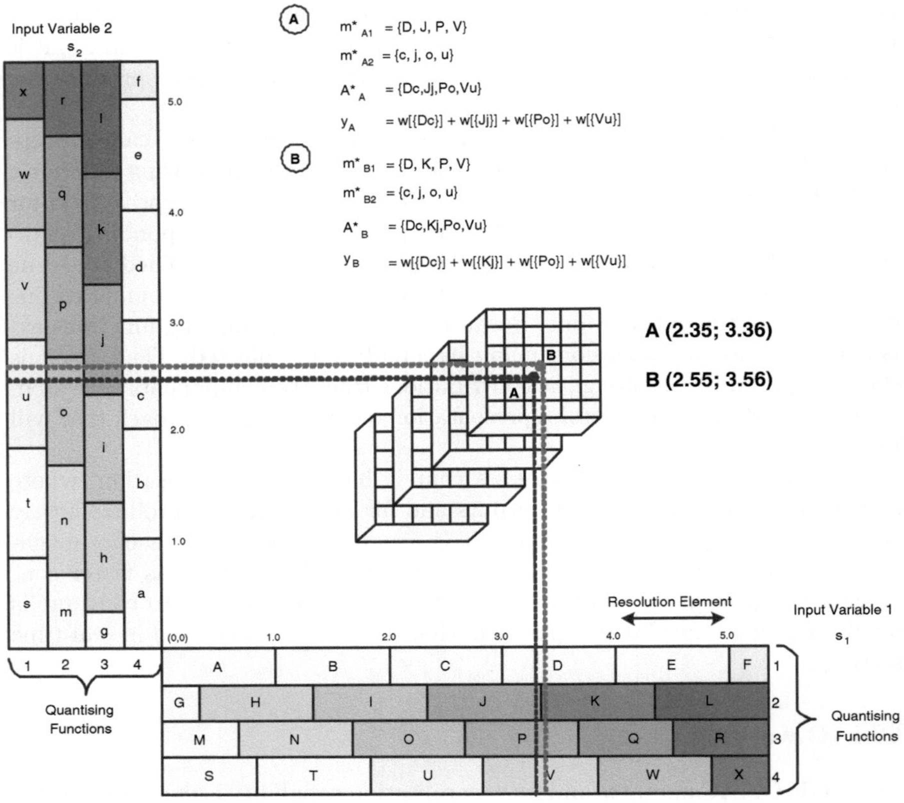

Fig. 7.3. Conceptual view of CMAC

The weights represent the values that map the selected hypercubes into the new weight space.

In the same way, $A^*_B = \{\{Dc\},\{Kj\},\{Po\},\{Vu\}\}$, and $y_B = w[\{Dc\}] + w[\{Kj\}] + w[\{Po\}] + w[\{Vu\}]$.

It can be noted that only one of the components of A^*_A and A^*_B differ. Hence, the output value y_B is similar to y_A, sice three out of four indexed weights that will be summed up are the same for both points. In fact, the weights for point B have not been previously updated, but due to the points similarity, the output for it can mostly be computed from the weights obtained from A^*_A.

CMAC is a linear mapping function from the input-output point of view, with a non-linear mapping executed between the input-output vectors. Each time an input vector is presented to compute an output, two mappings take place within the CMAC. The first one is responsible for transforming the input vector s in a binary-valued vector x with a higher dimensionality, through a non-linear fixed mapping $\phi(.)$, with $x = \phi(s)$, while the second, which is a linear mapping, multiply x for the current weights w in order to calculate y. The combination of these mappings is $g(.)$, where $y = g(\phi(s)) = g(x)$.

The mapping process from the input to the output is

$$\overbrace{s \rightarrow m^* \rightarrow A^* \rightarrow}^{\phi(.)} A_P^* \rightarrow y \tag{7.4}$$

where $\phi(.)$ consists of three first stages, with the other two representing the linear part of the mapping.

The set m^* is the set of receptive fields or resolution elements, which combined form the set of hypercubes, represented in A^*. These hypercubes index the weights that will be summed up to obtain the output. The mapping from A^* to A_p^* is done to reduce the memory storage necessity, with the later representing the actual space. The step $s \rightarrow A^*$ is called the generalisation step.

For the purposes of this paper, and to work properly with RL methods, the on-line approach of CMAC implementation has to be considered.

The pseudo code for Q-learning associated with CMAC is presented next.

```
Initialise Q(s,a) arbitrarily
Initialise A* with random numbers on
the range [0, N_w]
Initialise A*_p with zero values
Repeat (for each course)
      Initialise s
      Repeat for each trial
        Choose a ∈ A(s)
        Observe s_{t+1} and r
        For each tilling do
          Apply the quantising functions in each input
          Map A* to A*_p
          Q(s,a) ← Q(s,a) + α[r + γeQ(s_{t+1}) − Q(s,a)]
      s ← s_{t+1}
      until s reaches a terminal state
```

where $Q(s,a)$ is the y^* output desired value. For more details refer to [31].

7.4 Experiments

The proposed method of learning a switching policy was verified on a set of robot tasks and it was tested in a Khepera simulator environment. Two approaches for space discretisation were used, a feature-based and a state-based. The learned policies over these two models were compared to a predefined hand-crafted one.

7.4.1 Environment

The approach was tested in a Khepera robot simulator called YAKS [46]. From the set of sensors modelled by the simulator, the chosen ones were eight Infra-red sensors and eight light sensors; a vision turret sensor with a 1x64 vector of pixels with 256 grayscale levels, and a gripper with an object presence sensor. The world was defined as a 500mm X 500mm wall closed room, with different objects placed around it. The possible objects were: walls, dynamic objects (named sobst), static objects, balls and light sources. All objects had different grayscale level representation and specific constant radius. The environment was the same for all tasks and all policies applied.

7.4.2 Tasks

Six different tasks where proposed:

1. To find a sobst in the environment and align the robot to it
2. To find a sobst in the environment, align the robot and get close to it
3. To find a sobst in the environment and catch it
4. To find a sobst in the environment, catch it and displace it on an specific position
5. To find a sobst in the environment, catch it, find a light source, align to it and displace the object
6. To find a sobst in the environment, catch it, find a light source, go to its center and displace the object

7.4.3 Behaviours

A set of general and specific controllers were implemented. Each task has a set of actions (controllers) that can be chosen. The ones to be directly used by the tasks are:

1.safeWandering, 2.avoidBallCollision, 3.AlignToSobst, 4.goToSobst, 5.catchSobst, 6.dropSobst, 7.alignToLight 8.goToLight and 9.unStuck.

For each task, the action set A is:
Task1: actions 1, 2, 3 and 9
Task2: actions 1, 2, 3, 4 and 9
Task3: actions 1, 2, 3, 4, 5 and 9
Task4: actions 1, 2, 3, 4, 5, 6 and 9
Task5: actions 1, 2, 3, 4, 5, 6, 7 and 9
Task6: actions 1, 2, 3, 4, 5, 6, 7, 8 and 9

7.4.4 Space Discretisation

Two state space discretisations models were considered. The first, the feature-based, apply features as the basis for representing the state space, whereas the second, the state-based, uses the information directly provided by the robot sensors for representing the state space.

Feature-based

The feature model representation uses the concept presented in section 7.2.3 to define the tasks decomposition, to determine goals and subgoals, to find out which modules (controllers with associated operating conditions) should be built and how the features should be modelled. For details about the set of features and operating conditions encountered and the process of finding them out, see [47].

State-based

The state-based model for the tasks includes the following items in their state input vector:

Goal	stuck	seeBall	seeWall	seeSobst	seeObst	Ir[0-7]	Gripper	Light[0-7]

Each one can be either 0(active) or 1(non-active). Tasks 1 and 2 uses from Goal to Ir[7], task 3 and 4 uses until the Gripper and the last two tasks need the whole input vector represented. From this, the number of states for each task is: Task1/Task2: $2^2 * 2^4 * 2^8 = 16384$ states, Task3/Task4: $2^2 * 2^4 * 2^8 * 2^1 = 32768$ states and Task5/Task6: $2^2 * 2^4 * 2^8 * 2^1 * 2^8 = 8388608$ states.

7.4.5 Hand-crafted Policy

The hand-crafted policy, differently from the learned one, does not find a switching function, but instead applies a pre-defined sequence of actions based on the robot input vector (actual state). For more details see [47].

7.4.6 Learned Policies

The learned policies refer to the application of the Q-learning algorithm as the method for learning the switching function. All the experiments were carried out in the same environment (world size and objects displacement). For each task, over each discretisation model (state or feature) the experiments were organised on courses and trials. Each course refers to the complete execution of a set of trials, where a trial ran until the robot reached the goal or failed to do it in a predefined maximum number of steps. For the first four tasks, the number of maximum steps was empirically stablished as 150 and for the last two tasks it was established as 250. The number of courses is 20 and the number of trials is 500.

The Q-learning parameters are: $\gamma = 0.99$, $\alpha = 0.9$ and the initial exploration rate = 0.9, decaying with time to zero. The cost structure followed the principle that dense rewards can facilitate exploration. So, the cost of using each behaviour is one, except when the goal is reached, when a zero is received instead. In some situations, it can occur that the robot becomes stuck.

To prevent this state, a cost of $\frac{1}{1-\gamma}$ (equivalent to never reaching the goal) is communicated to the learning agent.

Performance of the algorithm was evaluated by the number of steps to reach the goal. Therefore, the objective of Q-learning was to minimise the number of steps needed to accomplish the task.

7.4.7 Results

After defining the number of steps each trial could last, the Q-learning algorithm was applied for each task over the two state space models. Their results were compared to the hand-crafted policy executed before.

Task 1. In figure 7.4 it can be noticed that both learned policies achieve better results than the hand-crafted one in only few trials. Figure 7.5 shows that, due to the exploration factor present on the algorithm and to its trial and error nature, the learning policies switch behaviours much more often than the hand-crafted one, that has a very tight sequence of actions.

It was also noticed that the state-based policy had better performance than the feature-based.

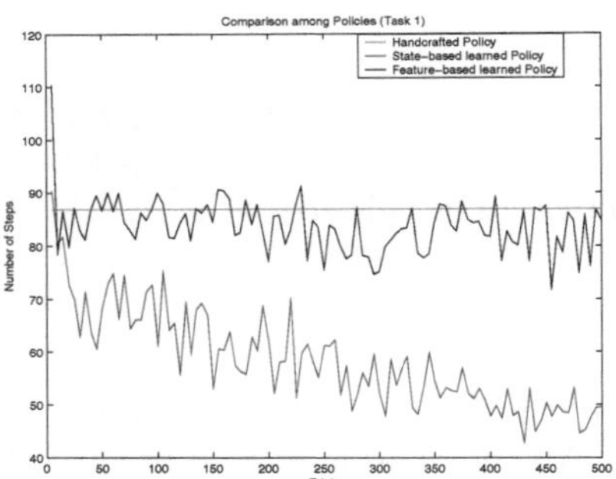

Fig. 7.4. Policies comparison for Task 1 - Number of steps

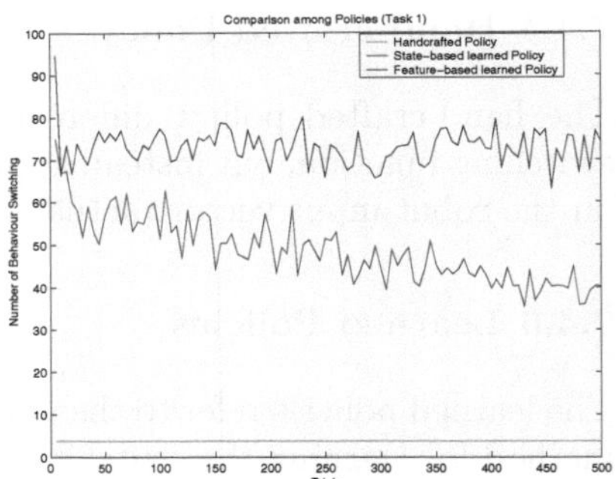

Fig. 7.5. Policies comparison for Task 1 - Behaviour switching

Task 2. For a slightly more complex task, Task 2, figures 7.6 and 7.7 lead us to the same evaluation and conclusions we arrived for the first executed task.

Task 3. In Task 3, as one can notice in figures 7.8 and 7.9 a slight change of roles on the quality of the learned policies happened. Now, the feature-based policy achieved a better result than the state-based one. This change happened because the system complexity (number of actions, states and task

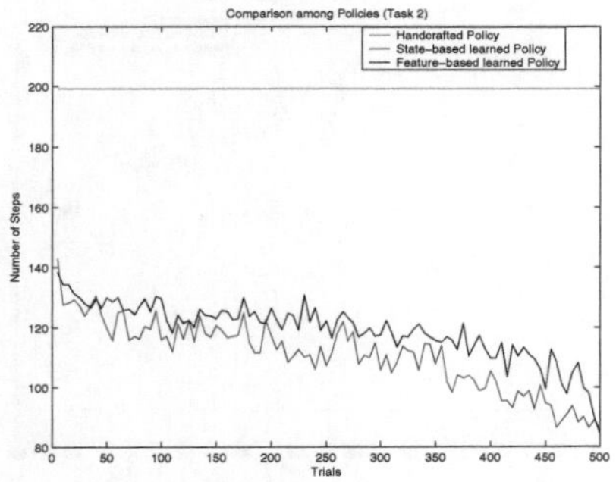

Fig. 7.6. Policies comparison for Task 2 - Number of steps

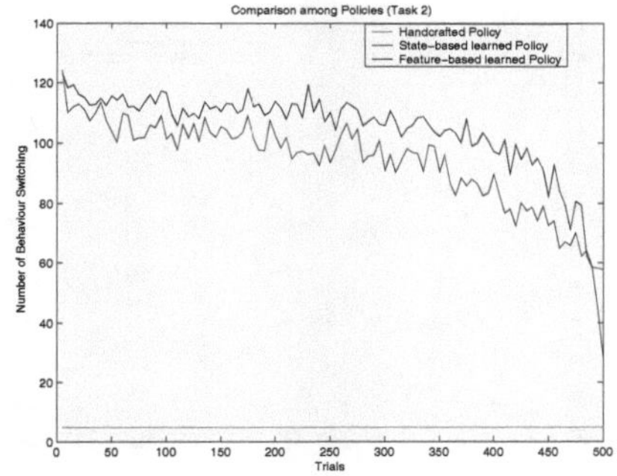

Fig. 7.7. Policies comparison for Task 2 - Behaviour switching

difficulty) has increased. As we saw in section 7.3, for convergence, it is necessary that all state-action pairs are visited infinitely often. In fact, it is much easier to explicitly visit 32.768 (the number of features-action pairs) state-action locations than 196.608 (the number of state-action pairs).

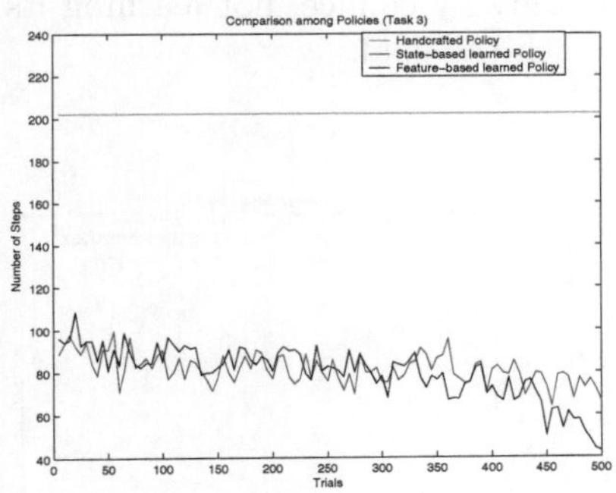

Fig. 7.8. Policies comparison for Task 3 - Number of steps

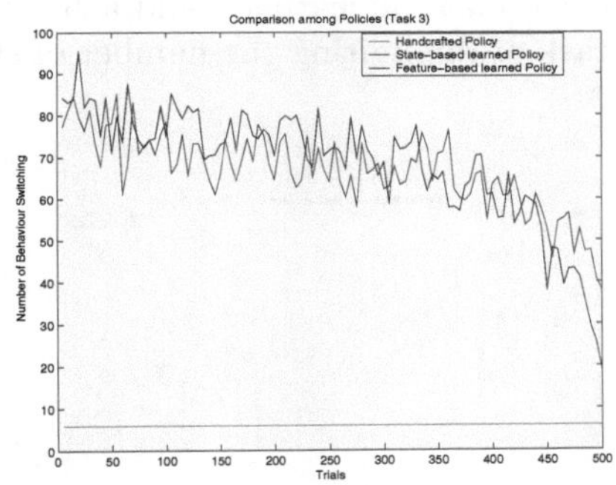

Fig. 7.9. Policies comparison for Task 3 - Behaviour switching

Task 4. For Task 4 (see Figures 7.10 and 7.11), we arrived to the same conclusions as we did for Task 3, i.e., the feature-based policy presents a faster convergence and better values than the state-based one. As was also noticed before for the first three tasks, the performance of any of the learned policies is far better than the hand-crafted one.

Task 5. Figure 7.12 presents the result for the execution of the feature-based learned strategy. As encountered before, the results were better than that for the hand-crafted policy that took, on average, 512 steps to execute the task. However, the number of successful trials decreased (figure 7.13).

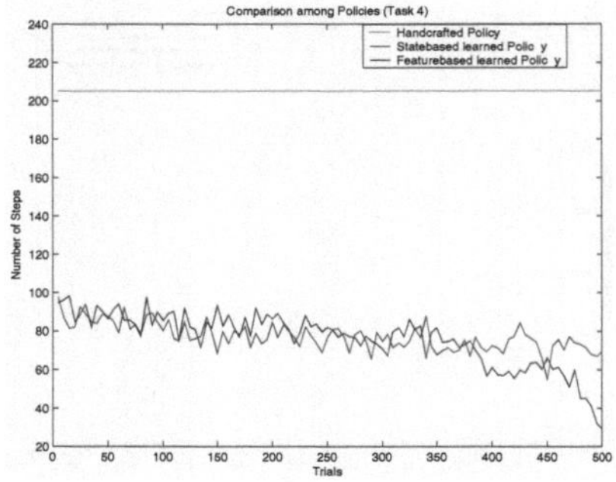

Fig. 7.10. Policies comparison for Task 4 - Number of steps

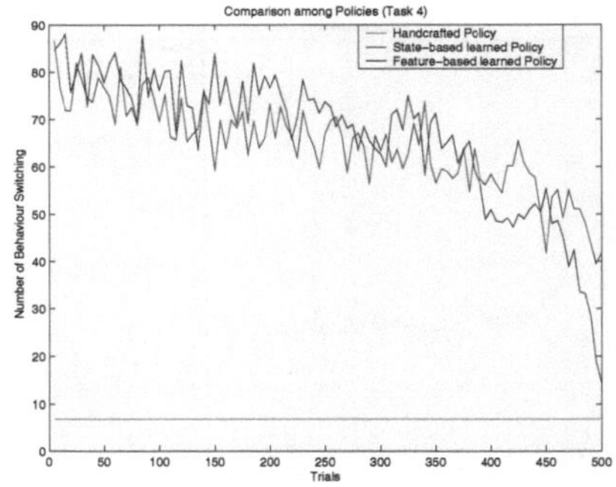

Fig. 7.11. Policies comparison for Task 4 - Behaviour switching

Unfortunately, for task 5, the space representation using states increased too much with the addition of the light sensors representation, thus compromising the learning task. While, for the feature representation, the total number of state-action locations increased by 4.57 the space needed for task 4, for the state representation we found an increasing rate of 292.

Figure 7.14 shows the state-based policy executing for task 5, where it is clear that the learner could achieve the goal only by chance, not learning its task of minimising the number of steps to reach the goal.

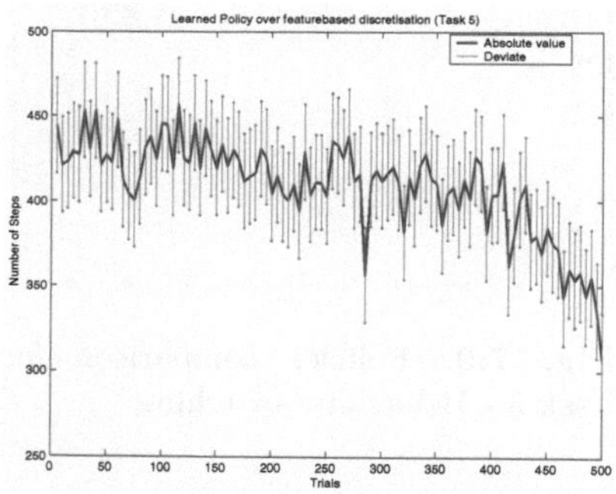

Fig. 7.12. Policies comparison for Task 5 - Number of steps

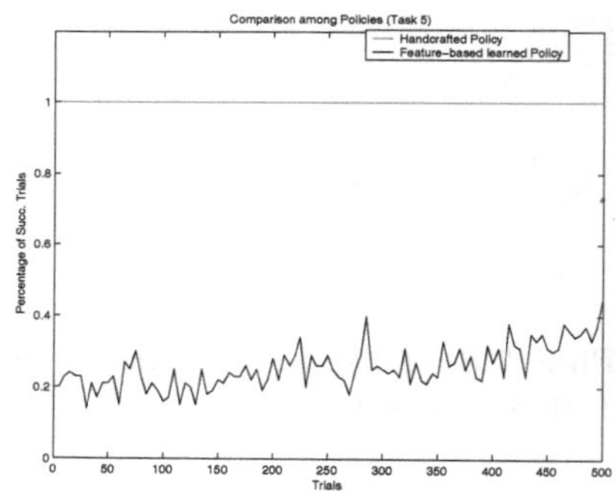

Fig. 7.13. Policies comparison for Task 5 - Percentage of successful trials

Task 6. For Task 6 (Figures 7.15 and 7.16) we found the same results as for Task 5, with an even worse performance for the state-based representation, as will be discussed next.

One can argue that the system could not be able to find a solution (not even by trial-and-error) because the upper-bound for the steps was set low. Actually, this could be a reasonable point and the number could be increased.

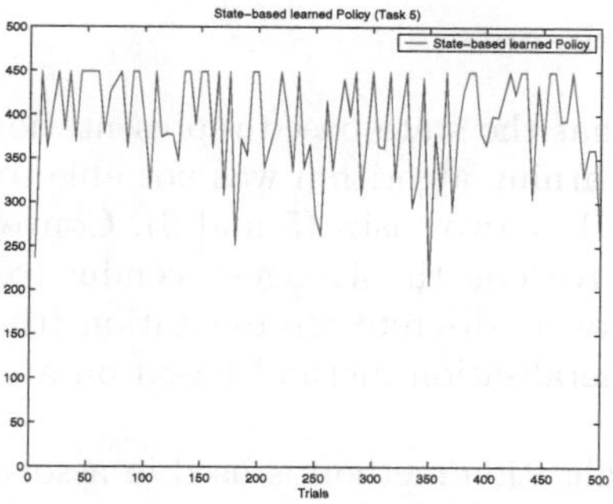

Fig. 7.14. State-based learned policy - Task 5

However, for the limit established, the total experiment lasted 960 hours (working on a Pentium4, 2.0GHz, 512MB RAM), what is obviously too much for an algorithm that should be executed in an autonomous mobile robot.

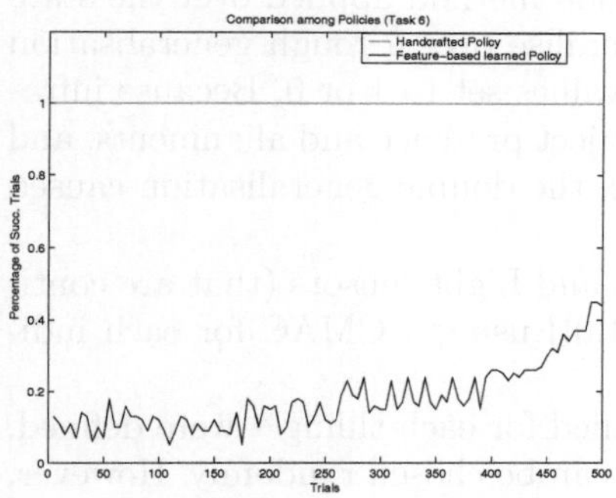

Fig. 7.15. Policies comparison for Task 6 - Number of steps

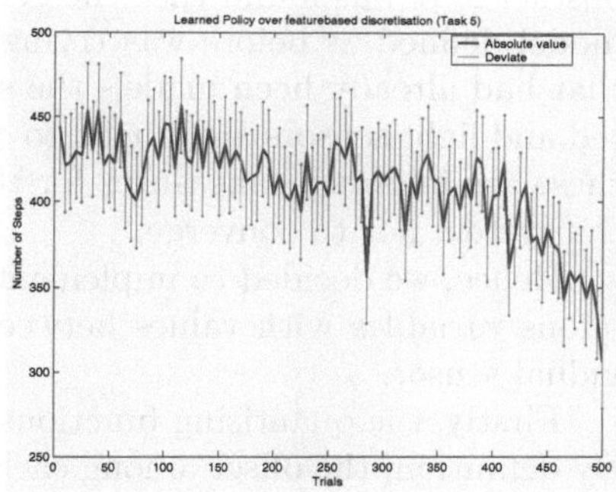

Fig. 7.16. Policies comparison for Task 6 - Percentage of successful trials

Table 7.1 presents a summary of the experiments carried out until this point. The values represented are average values calculated over the whole experiment.

Table 7.1. Average Number of Steps to Achieve the Goal per Task

	Average Number of Steps		
	Feature-based	State-based	Handcrafted
Task 1	83.71	58.38	86.8
Task 2	119.13	110.82	199.2
Task 3	80.01	81.48	202.2
Task 4	76.12	77.24	205.2
Task 5	409.20	-	512
Task 6	509.66	-	537

7.4.8 CMAC Experiments

So far, it has shown that the state-based representation worked well to tasks 1-4, although the Q-learning algorithm was not able to learn over this model representation for the last two tasks (5 and 6). Convergence could have not happened for different reasons, the strongest seeming to be the unfeasibly large state space achieved for its discrete representation (that contains $2^{23} states$). For this reason, a generalisation method based on a compact CMAC representation is proposed.

The CMAC generalisation method is used in association with Q-learning, to learn a switching action policy for the specified task.

CMAC Applied to State-based Representation

At the beginning, we tried to implement the CMAC over the state-based model with the representation it has been used until now - each sensor represented by 0 or 1 values. However, we did not succeed, even for the simple task of avoiding obstacles. It was found that the generalisation method applied over the state model defined as before was trying to generalise over a rough generalisation that had already been made - the sensors values set to 1 or 0. Because infrared and light sensors are crucial to define object presence and alignments, and these are important measures for the tasks, the double generalisation caused the system not to converge.

Hence, we decided to implement the IR and Light sensors (that are continuous variables with values between $[0.0, 1.0]$) using a CMAC for each individual sensor.

Firstly, the quantising functions – specified for each tilling – were defined. By definition, the offset among each tilling can be chosen randomly. However, in this work, a constant 0.05 value was applied.

The quantising function that presented the best results is shown in Figure 7.17. It is a non-uniform representation that incorporates the fact that values that indicate an obstacle close to the robot have to be more precise than those which represent this obstacle far away from it (High sensor values represent the obstacle very close to the robot).

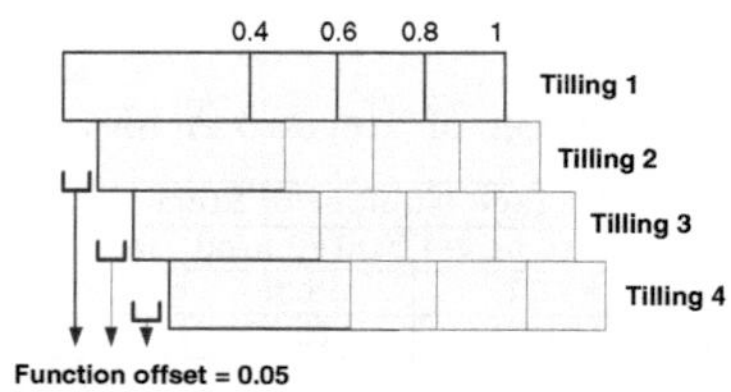

Fig. 7.17. CMAC discretisation for IR sensors

To apply the CMAC to execute tasks 5 and 6, it was necessary, after using it to each individual sensor, to define an spatial connection among them. This contrasts with space coordinates, where, whatever the number of dimensions is, there is a natural link among the values of each dimension. What means that, if we are using a spatial representation of states, once CMAC was applied to each individual dimension, a relation among the values would emerge naturally on the resultant hypercube.

As for sensors values, this natural link does not exist. Hence, one has to introduce, after applying CMAC individually to each sensor, another generalisation over specific pair of sensors we expect to have some relationship.

The sensors distribution in Figure 7.18 shows each sensor with CMAC working individually. After this step (first generalisation), each pair of sensors ([0 and 1], [1 and 2], [2 and 3], [3 and 4], [4 and 5] and [6 and 7]) had another CMAC working on its outputs. It consists of a different quantising function with 4 resolution elements to achieve generalisation. The red and green graphic in Figure 7.18 shows two individual sensors with their first-generalised values (in red). This values are then put together and generalised to another vector (in green) whose Q-values related are those that will be processed, updated and stored after hashing.

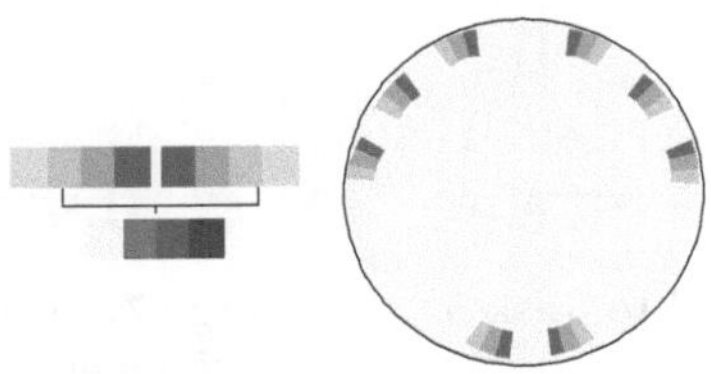

Fig. 7.18. CMAC resolution elements

Although the amount of memory required to manipulate the CMAC intermediate vectors is very high (2^{12}), the real memory (hash table) that persists the Q-values referent to state-action pairs, after both generalisations, and that has to be visited by the learning algorithm is about 100 (considering the 2^{23} states for tasks 5 and 6) times smaller than that required by a tabular representation. The size of the hash table was empirically established as 650.000.

For both the individual and pair CMAC application the number of tillings is 4, as well as the number of resolution elements for each quantising function for the individual sensors and for the pairs. The γ value is 0.7 and the experiment last for 500 trials.

The scheme was the same both for IR and Light sensors.

Figure 7.19 shows the result achieved for Task 5. It can be seen that although the feature-based representation has a better performance and converges to a better final result, the application of CMAC allows the the same task to be executed over a state representation where the Q-learning was not able to work at all. It also seems that CMAC convergence is slower than that of Q-learning applied to the feature-based model. For verification, one could require a bigger number of trials to verify the complete convergence of the algorithm. However, this could not be done because the experiment was already long enough (lasted about 9600 hours) and because the goal we had of showing CMAC working together with a Q-learning algorithm to reduce the state space complexity was achieved. In Figure 7.20 the result for Task 6 is shown. The application of CMAC allows the learning algorithm to achieve its goal under the state-based representation, although it is not capable of reducing the number of steps as much as the feature-based learned policy does.

Thus, the utilisation of CMAC allowed the use of a state-based model representation and lead to better results than the hand-crafted policy for both tasks.

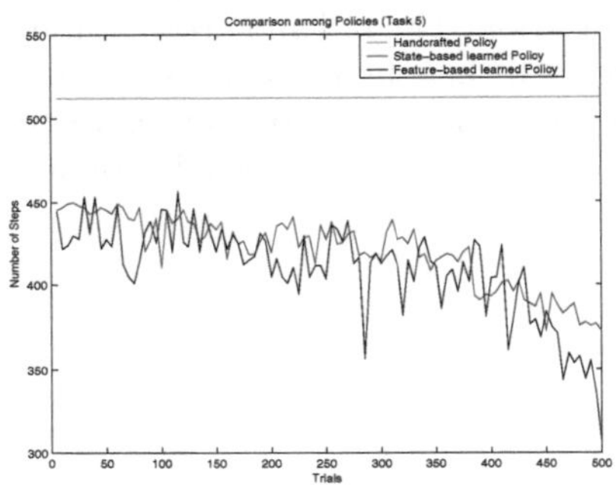

Fig. 7.19. CMAC applied to Task 5 - Number of Steps

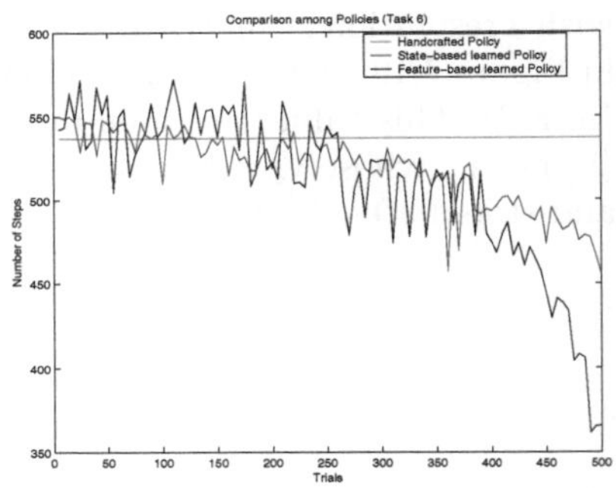

Fig. 7.20. CMAC applied to Task 6 - Number of Steps

7.5 Conclusions

The goal of the experiments described in this work was to evaluate the behaviour of the reinforcement learning algorithm Q-learning when applied to different state space discretisations in robot tasks.

For this purpose, two different space discretisations were applied. Both used concepts (abstractions, macro-operators and subgoals) present on the literature and the a priori knowledge about the environment and tasks to reduce the complexity of the problem and determine the set of controllers needed. The first discretisation model uses observation functions of the environment whereas the second is an almost direct representation of the sensors values as state description for the task at hand.

To allow the comparison of the above mentioned policies to a non-learned algorithm, a hard-wired sequence of actions - the hand-crafted policy - was implemented.

We found out that for the less complex tasks (1 to 4), the state-based representation presented better results than the feature-based learned policy. Furthermore, this state-based representation is easier to implement, once it only uses direct measurements of sensors. It was also noticed that both learned policies reached better results than those achieved by the hand-crafted policy.

However, in the course of the work we realised that the state-based discretisation could not be the used by the learning algorithm, for complex robotic tasks, as its space representation.

To continue the evaluation we intended at the beginning, we implemented the CMAC compact representation technique over the state-based space representation for the most complex defined tasks (5 and 6).

For these cases, we found out that the CMAC representation is well fitted for the tasks it were applied, although it only presents better results than the learned policy that was not able to learn at all. When compared to the feature-based learned policy results its values were not as good. It was also verified that this representation is easier to implement over spatially defined values, i.e., values that have intrinsic spatial relationship among themselves. In these cases the representation can be easily extended to n-dimensional spaces.

The CMAC takes also less steps to converge than the unfeasible state-based learned policy, but takes much more computational time to be executed, due to the mappings that are executed inside CMAC.

When applied additional sensor noise, the results showed that the values achieved by the learned policy over the more detailed representation (state-based) are more affected than those of the feature-based, what lead us to think the last as a more robust definition for a state space.

Finally, we realised that adaptive algorithms can be advantageous over non-adaptive techniques when applied to complex environments and tasks, thus extending the results previously presented in [22]. In particular, it seems that the conservatism of the hand-crafted policy did not allow it to exploit certain properties of the environment that could not be easily foreseen before its implementation.

References

1. Brooks, R. A. New Approaches to Robotics. In: Science, 1991. v. 253, p.1227-1232.
2. Sutton, R. S. Temporal Credit Assignement in Reinforcement Learning. PhD Thesis - University of Massachusetts, The MIT Press, 1984.
3. Sutton, R. S.; Barto, A. Reinforcement Learning: an Introduction. The MIT Press, 1998.
4. Werbös, P. J. Advanced Forecasting Methods for Global Crisis Warning and Models of Intelligence. In: General Systems Yearbook, 1977. v. 22, p. 25 - 38.
5. Bellman, R. Dynamic Programming. Princeton, New Jersey: Princeton University Press, 1957.

6. Bertsekas, D. P. A Counterexample to Temporal Differences Learning. In: Neural Computation, 1995. v. 7, p. 270 - 279.
7. Bertsekas, D. P. Dynamic Programming and Optimal Control. Massachusetts: Athena Scientific Belmont, 1995.
8. Littman, M. L.; Szepesvári, G. A generalized Reinforcement-learning Model: Convergence and Applications. In: Proceedings of the 13th International Conference on Machine Learning. Bari, Italy: Morgan Kaufmann, 1996. p. 310 - 318.
9. Pólya, G. How to Solve It? Princeton, NJ: Princeton University Press, 1945.
10. Newell, A.; Simon, H. A. Human Problem Solving. Englewood Cliffs, NJ: Prentice Hall, 1972.
11. Korf, R. E. Macro operators: A weak method for learning. In: Artificial Intelligence, 1985. v. 26, p. 35 - 77.
12. Korf, R. E. Planning as search: A quantitative approach. In: Artificial Intelligence, 1987. v. 33, p. 65 - 88.
13. Mahadevan, S.; CONNELL, J. Automatic programming of behavior-based robots using reinforcement learning. In: Artificial Intelligence, 1992. v. 55, p. 311 - 365.
14. Mataric, M. Reinforcement learning in the multi-robot domain. In: Autonomous Robot, 1997. v. 4.
15. Dorigo, M.; Colombetti, M. Robot shaping: Developing autonomous agents through learning. In: Artificial Intelligence, 1994. v. 71, p. 321 - 370.
16. Thrun, S.; Schwartz, A. Finding structure in reinforcement learning. In: Advances in Neural Information Processing Systems. Cambridge: The MIT Press, 1995. v. 7, p. 385 - 392.
17. Brockett, R. W. Hybrid models for motion control systems. In: Essays in Control: Perspectives in the Theory and its Applications. Boston: [s.n.], 1993. p. 29 - 53.
18. Sastry, S. Algorithms for design of hybrid systems. In: Proc. of the International Conference of Information Sciences. [S.l.: s.n.], 1997.
19. Maes, P. A bottom-up mechanism for behavior selection in an artificial creature. In: First International Conference on Simulation of Adaptive Behavior. MIT Press, 1991.
20. Maes, P.; Brooks, R. A. Learning to coordinate behaviors. In: National Conference on Artificial Intelligence, 1990. p. 796 - 802.
21. Brooks, R. A. Elephants don't play chess. Robotics and Autonomous Systems, v. 6, n. 1, p. 3-15, jun 1990.
22. Kalmar, Z.; Szepesvari, C.; Lorincz, A. Module-based reinforcement learning: Experiments with a real robot. Machine Learning, v. 31, n. 1-3, p. 55 - 85, April 1997.
23. Tsitsiklis, J. N.; Roy, B. V. Feature-based methods for large scale dynamic programming. In: Machine Learning, 1996. v. 22, p. 59-94.
24. Singh, S. P.; Jaakola, T.; JORDAN, M. I. Learning without state-estimation in partially observable markovian decision processes. In: Proc. of the Eleventh Machine Learning Conference, 1995. p. 284-292.
25. Lin, L. J. Reinforcement Learning for Robots using Neural Networks. PhD Thesis - Carnegie Mellon University, 1993.
26. Tesauro, G. Practical Issues in Temporal Difference Learning, 1991.
27. Boyan, J. A. Modular Neural Networks for Learning Context-dependent Game Strategies. Master's Thesis - Cambridge University, 1992.

28. Farley, B.; Clark, W. A. Simulation of self-organizing systems by digital computer. IRE Transactions on Information Theory, v. 4, p. 76-84, 1954.
29. Samuel, A. L. Some studies in machine learning using the game of checkers. IBM Journal on Research and Development, v. 3, p. 211-229, 1959.
30. Albus, J. S. Data storage in the cerebellar model articulation controller (cmac). Journal of Dynamic Systems Measurement and Control, v. 97, p. 228-233, 1975.
31. Watkins, C. Learning from Delayed Rewards. PhD Thesis - King's College - Cambridge, 1998.
32. Miller, W. T.; Kraft, L. G. Cmac: an Associative Neural Network Alternative to Backpropagation. In: IEEE Proceedings, 1990. v. 78, p. 1561-1567.
33. Kraft, L. G.; Campagna, D. P. A summary comparison of cmac neural network and traditional adaptive control systems. Neural Networks for Control, v. 3, p. 143-169, 1959.
34. Sutton, R. S. Generalization in reinforcement learning: Successful examples using sparse coarse coding. Advances in Neural Information Processing Systems, v. 8, p. 1038-1044, 1996.
35. Tham, C. K. Modular On-Line Function Approximation For Scaling Up Reinforcement Learning. PhD Thesis - Jesus College, Cambridge, England, 1994.
36. ESTEVES, C. H.; GABRIELLI, L.; RIBEIRO, C. Discretização cmac baseada em conhecimento a priori para generalização da experiência. In: Workshop sobre Informática na Escola, 2004.
37. Singh, S. P. et al. On the convergence of single-step on-policy reinforcement learning algorithms. In: Machine Learning, 1997.
38. Whitehead, S. D.; Ballard, D. H. Learning to perceive and act by trial and error. In: Machine Learning, 1992. v. 8, p. 3-4.
39. Mitchell, T. M. Machine Learning. Singapore: McGraw-Hill, 1997.
40. Koenig, S.; Simmons, R. G. Complexity analysis of real-time reinforcement learning. In: National Conference on Artificial Intelligence, 1993. p. 99-107.
41. Barto, A.; Sutton, R.; Anderson, C. Neuronlike elements that can solve difficult learning control problems. IEEE Transactions on Systems, Man and Cybernetics, SMC 13, p. 835-846, 1983.
42. Sutton, R. Integrated architectures for learning, planning, and reacting based on approximating dynamic programming. In: Proceedings of the Seventh International Conference on Machine Learning, 1990. p. 216-224.
43. Barto, A.; Bradtke, S.; Singh, S. Learning from Delayed Reinforcement in a Complex Domain, 1991.
44. Kim, H. Adaptive critic self-learning control. IEEE Transactions on Neural Networks, v. 2, n. 5, p. 530-532, 1991.
45. Albus, J. S. Data storage in the cerebellar model articulation controller (CMAC). Journal of Dynamic Systems Measurement and Control, v. 97, p. 228-233, 1975.
46. Carlsson, J. Yet Another Khepera Simulator (YAKS) homepage, 2001. Available at: ¡http://r2d2.ida.his.se¿. Acessed in: February-2005.
47. Colombini, E.L. Module-based Learning in Autonomous Mobile Robotics. Master's thesis, ITA, 2005.

8

A Hybrid Adaptive Architecture for Mobile Robots Based on Reactive Behaviours

Antonio Henrique Pinto Selvatici and Anna Helena Reali Costa

Laboratório de Técnicas Inteligentes – LTI
Escola Politécnica da Universidade de São Paulo
Av. Prof. Luciano Gualberto, trav.3, n.158
Cidade Universitária São Paulo - Brazil
{antonio.selvatici, anna.reali}@poli.usp.br

It is desirable that mobile robots applied to real world applications perform their tasks in previously unknown environments. Thus, a mobile robot architecture capable of adaptation is very suitable. This work presents a hybrid adaptive architecture for mobile robots called AAREACT that has the ability of learning how to coordinate primitive behaviors encoded by the Potential Fields method by using Reinforcement Learning. The proposed architecture is evaluated in terms of its performance curve when the robot is moved from one scenario to another. Experiments were performed on a Pioneer robot simulator, from ActivMedia Robotics®. Results suggest that AAREACT has good adaptation skills for specific environment and task.

8.1 Introduction

Intelligent Mobile Robots are physical agents that perform their tasks autonomously in the real world. Their actuation is determined from the processing of sensors input, which is done in cycles of alternate sensing (acquiring sensory information) and acting. The action to be executed is determined by the robot control system, which is an instance of a certain robotic architecture. The architecture can be considered as a framework for determining the robot actuation.

Architectures for mobile robots are usually designed following one of tree paradigms: reactive, hierarchical or hybrid deliberative/reactive [12]. The *reactive* paradigm consists in designing robots that determines its actions directly from the current sensory information, in a straightforward way. This task is usually decomposed in terms of reactive behaviors, which are modules that respond to the immediate sensory input. The final action results from a

A. H. P. Selvatici and A. H. R. Costa: *A Hybrid Adaptive Architecture for Mobile Robots Based on Reactive Behaviours*, Studies in Computational Intelligence (SCI) **50**, 161–184 (2007)
www.springerlink.com

simple coordination process among these behaviors. Thus, the robot resultant behavior emerges from its interaction with the world, which makes reactive architectures appropriate to cope with uncertainties about the environment. However, it is not guaranteed that the robot will accomplish its task, specially in complex environments. Reactive architectures are designed to perform well instantaneously, but are not capable of reasoning about the environment as a whole, planning actions for two or more further steps.

In the *hierarchical* paradigm, there is a planning stage between sensing and acting that decides the robot actions. The planning stage takes into consideration all the history of sensory information by building a world model, and performing some reasoning on it. Thus, the robot behavior derives from the world abstraction used to build the model. This allows the robot to plan its actions aiming at global optimality, but also ties its performance to the quality of the model used. If the abstraction adopted is not appropriate, or the world presents too many unexpected events, hierarchical architectures may lead the robot to fail in its objective.

The third paradigm, *hybrid deliberative/reactive*, is a way of combining the advantages of the other two. While reactive architectures are good in responding to an unpredictable environment, hierarchical ones can have a global vision of the robot situation, allowing the definition of actions aiming at a good overall performance. Hybrid architectures try to combine both advantages by means of a planning module that actuates over a set of reactive behaviors, sequencing their actuation, coordinating their outputs and/or modifying their internal mechanism.

Because of its greater complexity, the hybrid paradigm is usually used when pure reactive paradigm is not good enough for a satisfactory robot actuation [2]. It happens, for example, when an intelligent robot needs to adapt its architecture to changes in the environment in order to have a better performance. To adapt to new conditions, a robot architecture must have learning skills in order to observe and make criticisms to its own actuation, judging it based on some optimality measure, which is not possible for reactive architectures.

This work presents a hybrid architecture for mobile robots called AAREACT [16]. It consists of a coordination layer that learns the best way to combine the responses of reactive behaviors encoded by the potential fields method. These behaviors were inspired by the motor schemas architecture [2]. AAREACT can be seen as an architecture performing a hybrid control in a cooperative fashion: instead of selecting a single controller (behavior) each time, it determines the adequate influence of each behavior in the robot resultant action.

This chapter is organized as follows. Section 8.2 describes the high-level schemas commonly adopted for intelligent agent architectures, focusing on their constituent structural elements. Section 8.3 presents the main organization of the proposed architecture, which is detailed in sections 8.4 and 8.5, where the behaviors used and the coordination layer are respectively explained. Section 8.6 describes the experiments conducted with AAREACT and the results obtained. Section 8.7 presents some related work, inserting the architecture here presented into the context of other hybrid ones. The conclusions of this work are presented in section 8.8.

8.2 Agent Architectures

Considering mobile robots as autonomous agents, it is natural that the design of a robotic architecture should be preceded by the definition of which agent model the architecture will follow. This section discusses the agent classification proposed by Russel and Norvig [15]. Basically, the criterion adopted to classify agents considers the level of abstraction of the information employed to control their actions.

8.2.1 Simple Reflex Agents

Simple reflex agents are those which strictly follow the reactive paradigm. Actions are selected only on the basis of the current perception, ignoring all the past sensing history. The agent control law is implemented by means of simple rules directly linking the current perception with an action to be executed. They can be *condition-action* rules, expressed by **if-then** statements, or even mathematical formulae that express the action parameters as a function of sensory data. There is not any kind of planning or adaptation of the rules to be used. All the intentionality of the reflex-agents implicitly resides in the rules that produce the agent actions. There is no explicit intentionality: actions are the product of the interaction of the agent with its environment.

Although simple reflex agents can interpret sensory information in order to form a perception of the environment, they do not build any kind of model for it. All the history of perception is ignored, and then the agent cannot perceive how the world evolves. Thus, simple reflex agents can only infer from its sensors information in the abstraction level of events. Without the integration of the perception history, it is only possible for them to perceive eventual situations and react to them in a reflexive fashion.

8.2.2 Model-Based Reflex Agents

Simple reflexive agents have great difficulty in dealing with partial observations of the world. Not rarely, the success of an autonomous agent relies on its ability to recognize some aspects of the world that are not always visible to it. A more effective way to deal with world partial observability is controlling that part that cannot be perceived some times. To do so, the agent has to keep some kind of internal state that depends on the history of perceptions, reflecting one or more aspects of the world which is not observable every time.

The updating process of this internal information relies on two classes of knowledge, which must be encoded in the agent. The first of them is the knowledge about how the world evolves independently of the agent actions. Additionally, the agent needs to know how its own actions affect the world. All this knowledge about the world's behavior is known as the model of the world, which allows current perception to be combined with the agent internal state to generate the updated description of the world.

The interesting thing about this approach is that the model-based agent has the ability to retrieve not only eventual information from sensors data, but also information about more lasting situations. Besides the perception about the world current conditions, it is possible to make abstractions about how the world is. For example, if the agent notices the presence of an obstacle for many times when it passes through some place, the obstacle can be incorporated into the world model. Thus, persistent events can be abstracted to define a characteristic for the environment.

One more point to remark is that model-based reflex agents do not perfectly fit in the reactive paradigm, once the construction of a world representation is considered as a kind of deliberation. However, the control is performed in the same reflexive manner. The main difference is that perceptual information is retrieved from the world model internally represented, which is more complete than the perception that would be possible to extract only from the current sensory input. The interface that extracts information from the internal world representation is called virtual sensor [12].

8.2.3 Goal-Based Agents

If the internal representation of the world is a solution to the partial observability problem, reflexive agents also suffer from lack of flexibility in dealing with changes in the world. Once their programming is done by means of simple rules defined for a limited set of environmental situations and a specific task, a strong change in these elements can lead the agent to failure. Thus, it is desirable that the knowledge about the world be attached to some information describing desired situations, which the agent can combine with the knowledge about the results of its own actions in order to choose those that attain the goal.

Many times the goal pursued by the agent cannot be immediately attained from within one action, but different sequences of actions have to be considered to find a way to accomplish its objective. In these cases, the agent must have decision-making skills, which involves the implementation of search and planning techniques in order to provide reasoning about future possible situations.

The concept of goal represents an abstraction level superior to that of reactive rules. Goal-based agents have explicit intentionality, and the master regent of its actions is the goal, rather than the interaction with the environment. In this case, actions are steps towards the goal.

8.2.4 Utility-Based Agents

Although the explicit representation of goals provides the agent with a high level of flexibility to decide its actions, goal-based agents still present limited autonomy. The objectives need to be defined by the programmer, so that the decisions made by the goal-based agent do not represent an expression of "self-will". A greater level of autonomy would be achieved if the agent itself determined which goal is convenient at a given moment. To do so, information with a higher level of abstraction has to be employed to control the agent regarding something related to its "self-will". This can be achieved by means of a more general performance measure that permits a comparison of the possible "satisfaction level" among different goal states, if attained. This measure corresponds to a utility function, which maps a state or a sequence of states of the environment into a real number that expresses the associated "satisfaction level".

A complete specification of the utility function allows autonomous decisions to be made in two cases, for which goal specification only is not enough. First, when there are two contradictory reachable goals, the utility function specifies the most adequate one. The other case happens when agent aims at various goals, but none of them is guaranteed to be attained. In this case, the utility function provides a mean of pondering the probability of success with regard to the importance of the goals.

8.2.5 Learning Agents

Despite the high complexity presented by utility-based agents, there is an important skill of rational beings not addressed by any of the former agent models: learning. A learning agent can modify its own control laws in order to adapt to environmental changes and better accomplish its objectives.

However, implementing learning in artificial agents requires the definition of the information interchange structure, specifying the parameters to be adjusted and those variables that will be measured for expressing performance. To do so, the learning agent architecture can be divided into four conceptual components. The most important distinction to be done is between the *learning element*, responsible for the learning process, and the *performance element*, responsible for determining the actions to be executed by the agent. The performance element is equivalent to a complete agent architecture, following one of the four models previously discussed. So, the learning element adjusts the performance element parameters in order to improve the resultant actions.

The performance measure is a feedback about the agent behavior provided by the *critic*, which is compared to a desired pattern. The critic is necessary because the agent perceptions are not intended to offer clues about its degree of success. So, an independent observer is needed to inform the learning element about the performance evolution while the parameters are tuned.

The last component of the learning agent is the *problem generator*. It is responsible for suggesting actions that will lead to new and informative situations. The performance element itself would always behave in the same way for a given set of parameters defined by the learning element, which could be the best behavior regarding the acquired experience. In this case, the agent would be performing an exploitation of the acquired knowledge, but there would be no explicit action towards to improving the knowledge-base. However, the execution of some random actions, although not optimal in a short horizon, could imply the discover of a much better behavior when considering the long term performance.

One should notice that the learning agent model is formed by an embedded complete agent architecture equipped with a learning apparatus. This kind of embedding, although also present in other agent models, is more evident in this case. Thus, it is convenient to design a learning agent architecture on the basis of an existent well-studied one.

8.3 AAREACT

AAREACT is a robot architecture projected within the hybrid deliberative/reactive paradigm that combines adaptive skills and reactive behaviors. It is based on the concept of learning agents presented in section 8.2.5, which means that the architecture consists of a learning apparatus modifying parameters of an embedded sub-architecture based on the robot performance. The architecture schema is outlined in figure 8.1.

In AAREACT the embedded architecture corresponds to a simple reflex agent, called *reactive layer*. Despite the high complexity of the tasks that a more complete agent could accomplish, the pure reflexive agent can cope with several situations without the need for planning or other kind of deliberation. This leads to a simpler and faster implementation, operating in real time. Besides, reactive control is very appropriate in mobile robotics, once this approach was developed to overcome difficulties presented by robotic agents in their real-world operations.

The reactive layer is formed by reactive behaviors coordinated in a cooperative fashion. Each behavior is a module that indicates an action to be executed by the robot. Once these behaviors are reactive, they process sensors input in two main phases. Firstly, in the Sense phase, raw data is interpreted in a very straightforward way, resulting in perceptual information useful for the behavior. Then, in the Act phase, this information is transformed in an action command by a simple set of production rules or a mathematical formula. The final action command results from the combination of every behavior response performed by the Merging module. Thus, in contrast to the competitive coordination, which determines robot actions by just choosing one of those returned by the various behaviors, AAREACT reactive layer adopts

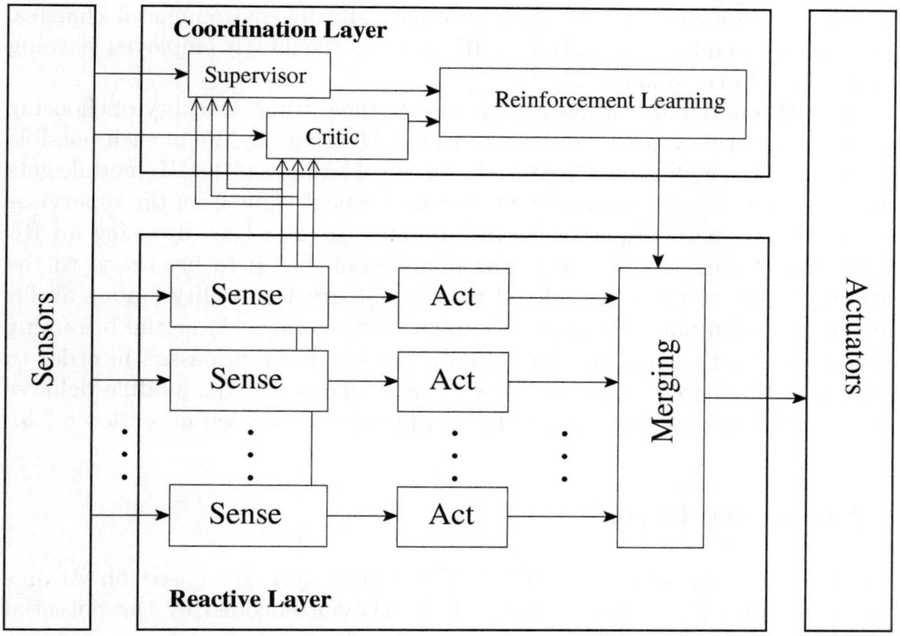

Fig. 8.1. AAREACT outline

cooperative coordination. However, the influences of each behavior response are adjusted by a pondering mechanism in the Merging module.

The learning part of the architecture is called *coordination layer*. Its role is that of adapting the value of the influence parameters that define the weight of each behavioral response in the resultant action. It is done regarding the situation of the environment, defined by a suitable interpretation of the robot sensors. Thus, the coordination layer has to learn the best way to coordinate the behaviors regarding the current sensory data.

The situation of the environment is determined by the supervisor module present in this layer, while the critic module observes the robot performance. One should notice that both modules may use raw sensor data and the perception information determined by behaviors perception. However, while the supervisor module considers only current information, the critic module has an internal state being updated during the interval of one situation. When the situation changes, the critic module defines a reinforcement value that summarizes how well it behaved while that situation was observed.

The learning model adopted in AAREACT is Reinforcement Learning (RL). RL is the problem faced by an agent that learns a policy of actuation based on interactions with a dynamic environment [8], being very suitable to model on-line untutored learning. Due to the autonomous nature of mobile robots, a non supervised technique is more suitable for robotic architectures, because it allows the robot adapt automatically to environmental changes, without the need for a tutor. In fact, RL is one of the mostly employed learning models in robotic systems [2].

The RL module maintains a data-base of the estimated utility of choosing a certain set of parameters to be sent to the Merging module in each possible situation of the environment. At each change of situation, the RL module gets the new situation information and the reinforcement value from the supervisor and critic modules, respectively, and updates its data-base by using an RL algorithm. Then, in most cases, the parameters chosen to be passed to the reactive layer are those considered to have the greatest utility among all for the current situation. However, as explained in section 8.5, at the beginning of the robot actuation, values randomly chosen are often passed in order to accelerate the learning process. When this happens, the RL module behaves as the problem generator present in learning agents, as seen in section 8.2.5.

8.4 Reactive Layer

The behaviors integrating AAREACT reactive layer are based on Arkin's motor schemas [2]. A motor schema is a behavior encoded by the potential fields method that translates the sensors readings into a movement vector directly and in a continuous fashion.

Each motor schema is composed by two modules: the perceptual schema and the encoding module. The perceptual schema is responsible for the Sense phase of the sensorial processing: it determines relevant information for controlling the robot from sensors input, thus generating the behavior perception. Then, the encoding module, which performs the Act phase, calculates a movement vector based on the potential fields method, using the perceptual information returned by the perceptual schema.

The potential fields method of behavioral encoding consists in generating a movement vector analogous to a force derived from some potential function, usually generated by associating repulsive charges to the obstacles and attractive charges to the target position. At each instant, each behavior calculates the forces generated by the interaction of the robot with the virtual potential field, and returns the resultant force as a movement vector. The final action command results from the vectorial sum of every vector returned. Depending on the architecture, this vector may have different meanings. It can be interpreted as indeed an external force pushing the robot, or just a velocity or displacement command. Here, the parameters that define the response of each behavior to the sensory stimulus are the speed and direction of motion.

In AAREACT, behaviors influences are adjusted before the calculation of the final action. The potential fields method suggests a very simple way to do it: the multiplication of each behavioral response by a pondering weight. Thus, in the Merging module, robot actions are determined by means of weighing the movement vectors returned by behaviors and then performing a vectorial sum of them. So, the role of the coordination layer of AAREACT turns to finding the best set of weights for each situation of the environment.

The behaviors that integrate AAREACT are:

avoidCollision

This behavior aims at avoiding the collision to the obstacles present in the environment. In its Sense phase, range readings are processed in order to identify the location of the detectable obstacles. Then the action vector is calculated analogously to the electrostatic force in Coulomb's law: repulsive charges are associated to each identified obstacle, generating move-away vectors with magnitudes that grow with the proximity to the obstacles. The behavior response is the vectorial sum of the calculated vectors for all detected obstacles. The equations that determine the movement parameters are:

$$V(d) = \begin{cases} V_{AC}\, e^{\frac{S-d}{T}} & \text{for } d > S \\ V_{AC} & \text{for } d \leq S \end{cases}, \qquad (8.1)$$
$$\phi = \pi - \phi_{\text{rob-obst}}$$

where V is the response magnitude (speed), d is the distance from the mass center of the robot to the obstacle, V_{AC} is the maximum speed allowed for the behavior, S is the robot stand off distance, T is the scale constant for the exponential function, ϕ is the motion direction and $\phi_{\text{rob-obst}}$ is the direction defined by the straight line that passes by the obstacle and the robot mass center.

moveToGoal

This behavior aims at attracting the robot to a pre-determined location in the environment. The target's position in the global coordinate system is informed by an external agent. The current robot position is determined by some localization method. The resultant motion direction is equal to the target's direction, given by the straight line that passes by the robot center and the target's location ($\phi_{\text{rob-targ}}$). The magnitude is given by a constant (V_{MTG}). Thus, the motion parameters' equations are:

$$\begin{aligned} V &= V_{MTG} \\ \phi &= \phi_{\text{rob-targ}} \end{aligned} . \qquad (8.2)$$

moveAhead

This behavior provides a certain trend for the robot not to change its heading direction. *moveAhead* does not use any sensory information, and the motion parameters are determined in a very simple manner: the magnitude is a constant (V_{MA}), and the direction is equal to the current robot heading (ϕ_{robot}):

$$\begin{aligned} V &= V_{MA} \\ \phi &= \phi_{\text{robot}} \end{aligned} . \qquad (8.3)$$

8.5 Coordination Layer

The coordination layer is responsible for learning the best policy for choosing the weights that multiply behavioral responses in the reactive layer. The learning process follows the approach of Module-Based Reinforcement Learning [9], in which a switching function implementing an RL algorithm chooses the best controller for the system regarding the current environment situation. Here, a controller is equivalent to a predefined set of weights that multiply behavioral responses, while the situation is defined by a vector of environmental features indicating the placement configuration of the obstacles and the target around the robot.

Figure 8.2 shows the complete AAREACT blocks diagram schema, detailing the information interchange among modules considering the reactive layer described in section 8.4. The learning module gets the information of the environment situation from the supervisor module, which determines it using the input from range sensors and the perception of *moveToGoal* behavior. The critic module defines rewards or penalties depending on the robot performance, measured on the basis of the approximation to the target and the average speed. Then, the learning module, implemented with SARSA algorithm[17], determines an adequate policy, that corresponds to choosing a set of weights for the behaviors according to the given situation. Thus, the robot coordination is defined by a set of three weights, w_{AC}, w_{MTG}, and w_{MA}, that respectively multiply the movement vectors returned by *avoidCollision*, *moveToGoal* and *moveAhead* behaviors.

8.5.1 SARSA Algorithm

Commonly, the role of RL algorithms is to determine, based on interactions with the environment, a function $Q(s, a)$, where Q is the utility of choosing action a given the environment situation s. In the case of AAREACT coordination layer, action a corresponds to the choice of a predefined set of weights.

When the agent observes a situation s and executes an action a, it obtains a response from the environment in terms of a reinforcement, r, that can be either a reward or a penalty. Then, a new situation s' is perceived. Once in this new situation, the agent has to choose a new action a'. SARSA [17] learning rule is defined from this 5-tuple $<s, a, r, s', a'>$:

$$Q(s, a) := Q(s, a) + \alpha(r + \gamma Q(s', a') - Q(s, a)). \tag{8.4}$$

where $r \in \mathbb{R}$ is the reinforcement signal, $\alpha \in]0, 1[$ is the learning rate, and $\gamma \in]0, 1[$ is the discount rate. Both α and γ are project parameters previously defined.

However, (8.4) does not provide a tool for determining the following action, which depends on the chosen strategy of actuation. If one decides to always trust in the current result of the learning process, the adopted strategy should

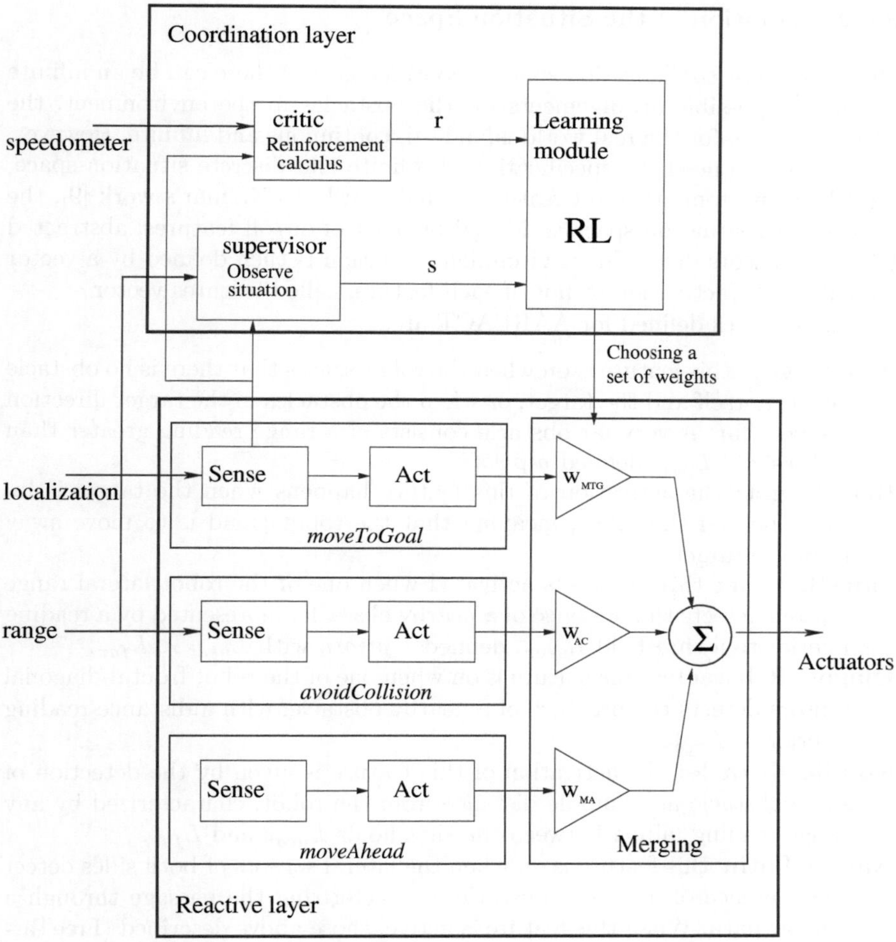

Fig. 8.2. AAREACT schema

be the greedy one, that always chooses the action to which is associated the greatest value of utility Q for the situation. However, a less confident strategy is more suitable to explore the possibility of actions, which is useful mainly in the initial phases of learning, when the agent is unlikely to have enough experience to decide the best action for the current situation. The strategy adopted in this work is the ϵ-*greedy* [17], that has the probability ϵ of choosing a random action, and a probability $1 - \epsilon$ of choosing the action with greatest associated utility.

8.5.2 Definition of the Situation Space

Once the robot configuration space is continuous, and there can be an infinite number of possible arrangements for the obstacles in the environment, the situation space for the real world is, indeed, continuous and infinite. However, RL algorithms need the specification of a finite and discrete situation space, modeling just some relevant features. Analogously to Kalmár's work [9], the environment situation space is defined by a set of on/off features, abstracted from the sensors data. The environment situation is then defined by a vector indicating the activation or not of each feature, called features vector.

The features defined for AAREACT are:

FreeTarget: this feature is on when the robot senses that there is no obstacle between itself and the target, or when the obstacles in the target direction are very far. A very far obstacle consists of a range reading greater than a threshold L_{far}, defined *a priori.*

BackTarget: the activation of this feature happens when the target is located behind the robot, meaning that the robot trend is to move away from the target.

SideObstacle: this feature is activated when one of the robot lateral range sensors detects the presence of a nearby obstacle, represented by a reading inferior to a threshold L_{near}, defined *a priori*, with $L_{near} < L_{far}$.

DiagonalObstacle: this feature is on when one of the robot frontal-diagonal sensors detects the presence of a nearby obstacle, with a distance reading inferior to L_{near}.

MiddleObstacle: the activation of this feature is given by the detection of some obstacle at a middle distance from the robot, characterized by any range reading falling between the thresholds L_{near} and L_{far}.

NarrowPath: this feature is on when the lateral sensors of both sides detect the presence of a nearby obstacle, characterizing the passage through a narrow path. When this feature is active, those above described (FreeTarget, BackTarget, SideObstacle, DiagonalObstacle and MiddleObstacle) are ignored, once we consider that the navigation in a narrow path is a special situation, when the robot should not take into consideration the target position or the presence of obstacles outside the path.

FrontalObstacle: this feature is active when one of the frontal sensors detects the presence of a nearby obstacle, characterizing danger of imminent collision. When this feature is on, all the others are ignored, once the robot priority reaction in this case should be avoiding this obstacle and not consider he target location or further obstacles.

Figure 8.3 illustrates the definition of each feature, showing the respective spacial distribution of obstacles and the target around the robot. The environment situation is defined by the features vector, signaling which features are on and which are off. A change of situation occurs when one or more inactive features are activated, or also when one or more activated features are deactivated. Given the number of predefined features, one can wrongly conclude that there are $2^7 = 128$ possible situations. However, considering the way features were defined, many of them are mutually excluding, which strongly reduces the number of possible situations. A further analysis demonstrates that only 24 situations are allowed.

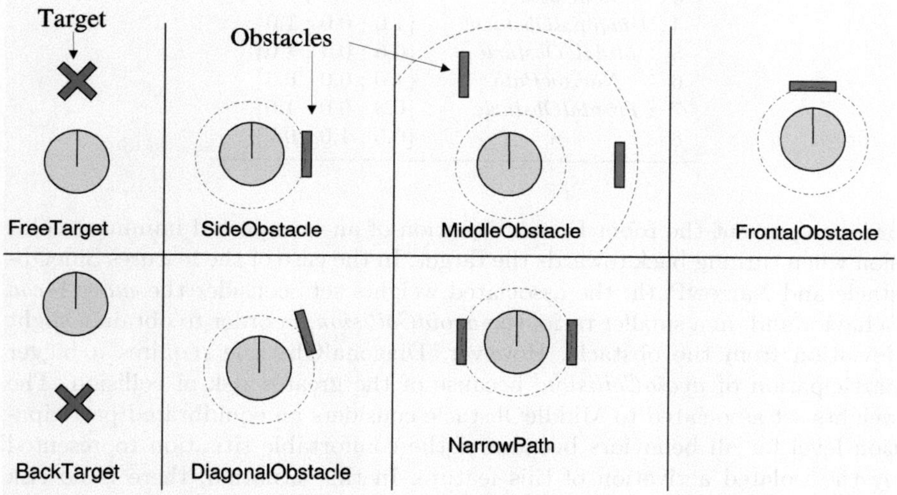

Fig. 8.3. Illustration of the defined features. In the picture, the robot is represented by the light gray circle, while the line segment inside it shows the direction of movement. The cross represents the target position, where the robot should move to, and the rectangles represent the detected obstacles. The dotted-trace circumferences represent the distances L_{near} and L_{far} (L_{near} is denoted by the circumference nearest to the robot).

8.5.3 Definition of the Weights Sets

The sets of weights were defined associating a controller to each defined feature. When the environment situation is defined by the activation of a *unique* feature, the chosen set of weights is that associated to the active feature. In the other situations, when more than one feature is on, the RL module inside the coordination layer has to choose a set of weights among those discriminated in table 8.1. The way the features were defined prevents having all of them off at the same time.

The weights set number 1 was associated to the feature FreeTarget, and considers only the *moveToGoal* behavior. Set number 2, associated to BackTarget, considers, in addition to *moveToGoal*, the *avoidCollision* behavior, in

Table 8.1. Defined features and the set of weights associated to each one.

Nb.	Feature	Associated weights set $\{w_{MA}; w_{MTG}; w_{AC}\}$
1	*FreeTarget*	{0.0 ; 1.0 ; 0.0}
2	*BackTarget*	{0.0 ; 1.0 ; 1.0}
3	*SideObstacle*	{1.0 ; 0.0 ; 0.3}
4	*DiagonalObstacle*	{1.0 ; 0.0 ; 1.0}
5	*MiddleObstacle*	{0.6 ; 0.4 ; 1.0}
6	*NarrowPath*	{1.0 ; 0.0 ; 0.3}
7	*FrontalObstacle*	{0.3 ; 0.0 ; 1.0}
8	—	{0.5 ; 1.0 ; 0.7}

order to prevent the robot from a situation of an unexpected imminent collision when turning back towards the target. In the case of the features SideObstacle and NarrowPath, the associated weights sets consider the *moveAhead* behavior and, in a smaller ratio, the *avoidCollision*, in order to obtain a slight deviation from the obstacle. However, DiagonalObstacle requires a bigger participation of *avoidCollision*, because of the greater risk of collision. The weights set associated to MiddleObstacle considers an equilibrated participation level for all behaviors because of the comfortable situation represented by the isolated activation of this feature. In this situation, there is no risk of imminent collision. Finally, the feature FrontalObstacle, which represents the danger of imminent collision, requires the maximum participation of the *avoidCollision* behavior, and a slight influence of *moveAhead* in order to avoid sharp robot movements.

One can notice that set number 8 is not associated to any feature. It was purposefully elaborated to be different enough from the others: other than set number 8, only set number 5 conjugates all the three behaviors simultaneously, but with different weights. Thus, the set of weight number 8 is present as an *additional option* to be explored in the case of activation of more than one feature at the same time.

8.5.4 The Reinforcement Function

In RL, positive reinforcements are used to reward desirable situations, while negative reinforcements can also be used to penalize undesirable situations. The robot primary objective is to get to the target's location. When this occurs, the critic module defines a great reward, defined by r_{goal}.

Besides, it is also desirable that the robot present a good performance during its actuation, which is difficult to express in numbers. While RL algorithms seeks to achieve a good overall long-term performance, short-term performance may be used to indicate wether the expected long-term one will be better or worse. In this work, two parameters of the robot trajectory were adopted to express its short-term performance: the average speed developed and the average speed of target approaching. When a situation change is detected, a reward proportional to these performance measures is calculated, characterizing an *intermediate reinforcement*, received before reaching the target. Intermediate reinforcements are useful for accelerating the learning process [11].

This reward, however, must be small when compared to r_{goal}. It is defined as

$$r_{int} = K_1 v_{av} + K_2 v_a, \tag{8.5}$$

where:

- K_1 and K_2 are gains arbitrarily defined;
- v_{av} is the average speed developed by the robot in the time interval of the duration of a situation, measured by a speedometer;
- v_a is the average velocity of target approaching performed during the time interval of the duration of a situation, given by the difference between the final and the initial distances from the robot to the target, divided by the time interval of the duration of the situation.

The above formulation allows positive and negative values for r_{int}. The value of this intermediate reinforcement is saturated in $\pm\bar{r}$, to warrant that $|r_{int}| < |r_{goal}|$.

8.6 Experiments with AAREACT

In order to verify the learning ability of the proposed architecture, AAREACT was implemented in a realistic robot simulator. Experiments were performed in two phases. In the first phase, named initial learning, the RL module does not have a good estimative of the utilities $Q(s, a)$, and a ϵ-greedy strategy is adopted. In this phase, the robot is expected to have a bad performance, even showing erratic trajectories. This kind of global behavior makes the robot prone to experimenting different situations initially, learning how to deal with them, rendering an effective learned policy that could be applied to different environment settings.

In the second phase, the knowledge acquired in the first phase is incorporated, so a better performance for the robot is expected. In this phase, the RL module does not explore the space of weight sets anymore, but exploits it according to the acquired knowledge by adopting a greedy strategy (ϵ is set to zero). It is worthy to emphasize that the learning algorithm keeps working as before.

8.6.1 The Robot Model

In the experiments, the robot model adopted was the Pioneer 2-DX, from ActivMedia® Robotics [1]. It has a 44cm long, 38cm wide and 22cm high aluminum body, weighing 9kg and capable of transporting another 23kg of additional load. The robot has a two-wheel drive plus a balancing caster wheel, with a differential steering. The robot can travel at a maximum speed of 1.6m/s, and its rotation rate can reach 230°/s.

The Pioneer sensors to be considered in this work are only those included in the basic package. Eight sonars, placed in a frontal semi-ring and working at a frequency of 25Hz, are the only range sensors used. The six sonars pointing forward are uniformly distributed within an angle of 90°, with a step angle of 15°between adjacent ones. Besides, there are two side sonars, one on each side, pointing to the left and the right. The robot localization is given by dead reckoning, using the 500-tick encoders present in each drive wheel. The speedometer is also implemented based on the encoders.

All the experiments were performed using the robot simulator, distributed together with the communication API. The sensors reading simulations are quite realistic, and the simulated errors behave very similarly to real ones. Therefore, the obtained results are expected to faithfully express those that would be achieved in a real scenario.

8.6.2 Initial Learning Phase

Once the knowledge acquired by AAREACT extends to diverse environments, it is interesting to take advantage of the initial learning to train the robot in an environment that presents a great variety of possible situations to be explored. The scenario projected for this first phase offers the robot several kinds of obstacles configuration before accomplishing its objective. This scenario consists of a rectangular room with four target positions near the corners, as shown in figure 8.4. The robot is then commanded to search one target at a time, ensuring that it will have circulated the room internally when it reaches the fourth target.

The AAREACT architecture is implemented with the following parameters:

- Behavior parameters
  - Stand off: $S = 40$cm
  - Exponential scaling constant: $T = 15$cm
  - moveAhead speed: $V_{MA} = 15.0$cm/s
  - avoidCollision speed: $V_{AC} = 15.0$cm/s
  - moveToGoal speed: $V_{MTG} = 15.0$cm/s
- Feature parameters:
  - Threshold for nearby obstacles: $L_{near} = 50$cm
  - Threshold for far obstacles: $L_{far} = 200$cm
- Learning algorithm parameters:
  - Learning rate: $\alpha = 0.3$
  - Discount rate: $\gamma = 0.99$
  - Probability of exploration: $\epsilon = 20\%$

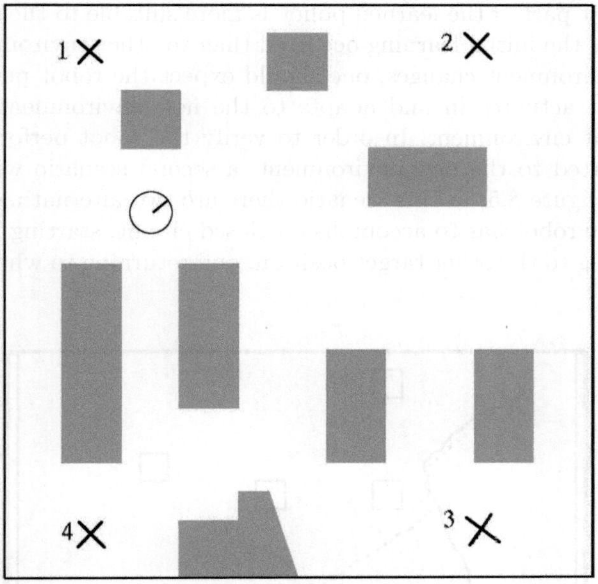

Fig. 8.4. Scenario were the simulated robot initiates its learning process. The positions marked with × are the targets locations, and the numbers next to them indicate the order they are sought. The elements in gray represent the obstacles, and the circumference with the internal line segment represent the robot and its respective direction of motion.

- Reward parameters:
  - Goal reward: $r_{goal} = 100$
  - Maximum intermediate reward: $\bar{r} = 3$
  - Gain over the average speed: $K_1 = 0.02\text{s/mm}$
  - Gain over the average velocity of target approaching: $K_2 = 0.002\text{s/mm}$

Since SARSA algorithm is always active, a simple stop criterion was defined for the initial learning: the first experiment was executed for the period of time necessary for the robot to show little variation of its long-term performance, which means that the acquired knowledge was good enough for the task accomplishment. The robot performance was measured by the time spent to complete the circuit at each epoch. The obtained Q values were used in the experiments of the second learning phase, when the probability of exploration ϵ is fixed to zero.

8.6.3 Scenario Changing Experiments

AAREACT concept was developed to allow the robot to show an efficient performance in any environment right after the initial learning phase. However, it is natural that part of the learned policy is more suitable to the specific environment were the initial learning occurred than to other environments. Thus, when the environment changes, one should expect the robot performance to improve as it actuates in and adapts to the new environment, tuning the policy to that environment. In order to verify the robot performance when it is transported to the new environment, a second scenario was projected. As shown in figure 8.5, in this scenario there are two alternating target positions, and the robot has to accomplish a closed circuit, starting in one target position, going to the other target position, and returning to where it started.

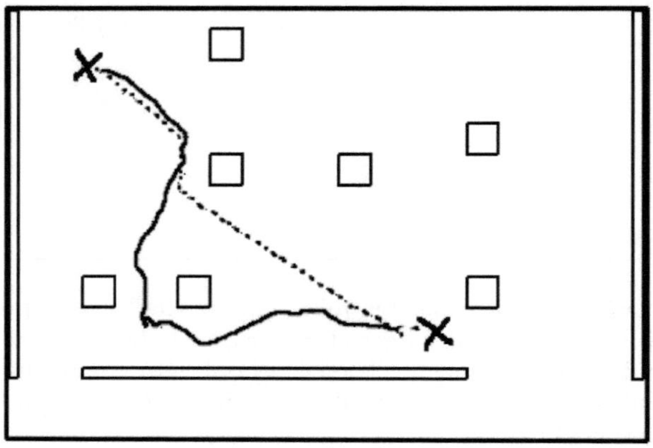

Fig. 8.5. Scenario for the second learning phase. The continuous trace represents a segment of the robot trajectory in the second learning epoch, and the dotted trace is a segment retired from the 30th. epoch.

One can observe that in the early learning epochs, AAREACT actuation in this scenario already showed a satisfactory performance, once the robot accomplished its objective in an acceptable time (less than 900s). However, the trajectories shown in figure 8.5 are a typical example of how the initial performance can be improved. In fact, as long as it actuated in the scenario, the robot performance tends to improve and stabilizes within a much smaller range (less then 300s), as shown in figure 8.6. While AAREACT is adapting to the specific scenario, its performance curve presents notorious variations due to the uncertainties generated by the interaction with the world.

8.7 Related Work

It is already common sense that practical robot architectures needs to follow the hybrid paradigm. Many hybrid architectures have already been proposed,

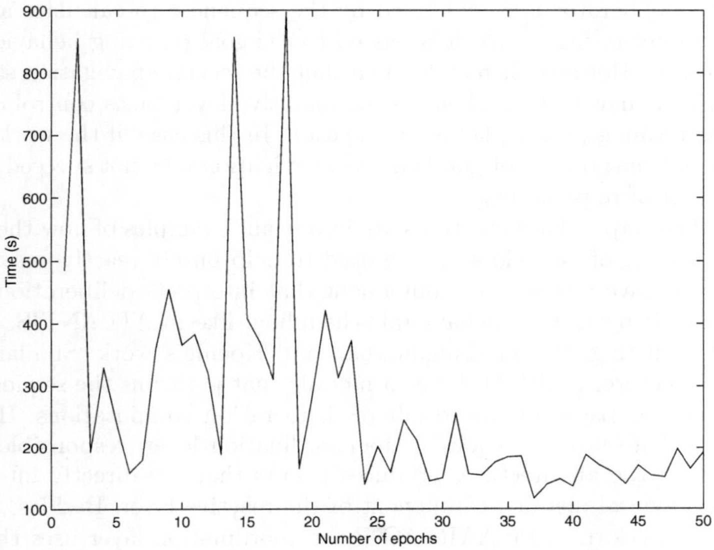

Fig. 8.6. Performance curve presented by AAREACT while adapting to the second scenario. It is the average curve obtained by repeating the experiment three times.

either for solving a particular problem or trying to consolidate a more general framework for the area. In most of them, the reactive part corresponds to modules responsible for generating action commands for the robot, while deliberation is used to define a policy for scheduling and sequencing these modules. This is the case of the so-called *three-layered architectures*, taken as examples to enforce the thought that purely deliberative navigation is not interesting [5]. An example of such architecture is ATLANTIS [4], which consists of three components. The *controller* is a reactive control mechanism responsible for primitive activities, performing no decision-making computations. The *deliberator* is responsible for time-consuming deliberative computations, like planning and maintaining world-models. The *sequencer* is like a special-purpose operating system, which makes the role of a manager for the architecture. It controls the initiation and termination of the activities performed by the controller and the deliberator, based on information provided by both components.

Another three-layered architecture was proposed by Ranganathan and Köenig [13]. Their architecture's components are: the *reactive layer*, responsible for the reactive navigation, the *deliberative layer*, which performs path planning and goal re-planning, and the *sequencing layer*, which decides whether the reactive or the deliberative layer should be active at the moment. The reactive layer is the one chosen in ordinary navigation, performed by reactive behaviors modeled by Arkin's motor schemas. If the robot gets into a situation of bad performance (mainly caused by local minima in the potential field), the deliberator is commanded by the sequencer to calculate an intermediate target position, which is passed to the goal pursuing behavior as its new objective. However, it may happen that the reactive navigation still fails, even with the new target. Then, the deliberative layer takes control over the robot, performing purely planned navigation. In this case, if the world model built for this purpose is not good enough, navigation may not succeed, or may require a lot of re-planning.

The three-layered architectures are important examples of how the deliberate sequencing of behaviors can be used to help purely reactive navigation. All of them have a sequencer component that interprets deliberation results and applies it by making behavioral scheduling, like in ATLANTIS, and behavioral adjusting, like in Ranganathan and Koenig's work. Similar to the later architecture, AAREACT has a module that performs the sequencing of motor schemas based on the result of deliberation computations. However, no intermediate layer is required: the coordination layer, responsible for deliberation in the architecture, produces results that are directly interpreted by the behavioral merging component of the reactive layer. Besides, no path planning is performed by AAREACT: the coordination layer uses the robot experience to learn the best way too coordinate the behavioral responses in order to obtain the resultant action, which is more aligned with the thought of avoiding purely planned navigation.

However, the actions of hybrid architectures are not exclusively produced by reactive controllers. In DAMN [14], an architecture developed for the specific task of road navigation, some of the behaviors are purely deliberative, using complex representation of the world. In that architecture, a more complex schema of behavioral coordination is performed. An arbiter decides on the robot actions based on each action command returned by behaviors, characterizing a cooperative coordination. The arbiter takes into consideration the weight of each behavior, constantly maintained by a deliberative mechanism.

More recently, the use of adaptive techniques as a form of deliberation has become a trend. An example is the architecture proposed in the work of Ishiguro and colleagues [7]. Like in many hybrid architectures, there are some behavioral modules coordinated in a competitive fashion. Their sequencing is determined on the basis of the results of deliberative computations, which take into consideration their scores, constantly updated using a simple adaptive algorithm. On the success of a certain behavior in accomplishing a task, the associated score is increased by a constant value. On the failure, this score is decreased by a proportional value. It is interesting to point out the presence of the learning agents elements in the architecture, regardless of the simplicity of the adaptive algorithm. In practice, autonomous adaptive architectures, despite the great differences among them, needs to follow this approach.

Another way of learning in hybrid architectures is behavior tuning. Although following the same learning agent model, the object of adaptation is the behavior mechanism rather than behavior coordination in this kind of architectures. In [10], a low level controller fuses action commands provided by two reactive behaviors: goal pursuing and obstacle avoidance. The controller function is to guarantee a smooth movement for the robot. The goal pursuing behavior is implemented by an Extended Kohonen Map which is trained to produce a sequence of motor velocity commands. The associated training data is the difference between the movement commanded by this behavior and the measured robot displacement, acquired after an action execution. Thus, the goal pursuing behavior is tuned in order to make the robot advance towards the target position as much as possible.

Despite the existence of other alternatives, the Reinforcement Learning model fits the autonomous agents learning case very well. Besides, because of its theoretical background, it provides very well funded techniques and consolidated tools. In robotics, a very common application of RL is the decision of robot actions based on the environment state. The work of Inoue and colleagues [6] is an example. The discrete state space corresponds to a quantization of the robot configuration space: the position space is divided into a homogeneous grid, and the possible orientation angles are quantized in 0°, 90°, 180°, and 270°. The possible action commands are only three: go forward 1 step, turn left, or turn right. The Q-Learning algorithm is used to determine the best action for each of the possible states. Because of the large state space size, the learning algorithm is likely to take a long time to converge to the best policy. To solve this problem, [6] proposes the use of a set of rules, obtained automatically by statistics techniques, in order to generalize learning results, reducing time of convergence.

However, defining the state and action spaces as simple discretizations of the continuous robot configuration and action spaces, respectively, brings serious drawbacks. An alternative is letting the state space be defined as a set of environmental features values, and action commands to correspond to predefined primitive behaviors. Then, the objective of the learning algorithm becomes finding an effective coordination policy. Kalmár and colleagues [9] studied this approach, naming it Module-Based Reinforcement Learningand applying it to the problem of learning behavior coordination for task-oriented mobile robots. In their work, the robot final objective is decomposed into some simpler subtasks, and then a specific behavior is designed to accomplish each subtask. The activation of any of them is flagged by the presence in the environment of the feature corresponding to the behavior operation condition. However, it occurs that sensory information may not be complete enough for the robot to accurately determine the adequate behavior to assume. It may happen that some evidences from the environment suggest the robot to adopt more than one behavior at a time. The approach used to solve this problem is the implementation of a central decider, that chooses the best behavior for a given situation based on the achieved experience. A RL technique is used to define the best policy for the choice of behaviors in each possible environment situation. The same architecture was also studied by Colombini and Ribeiro [3], using a very similar task to extend the results of [9]. Besides enforcing the conclusion that policies learned automatically are better than handcrafted ones, [3] shows that features extracted directly from sensors readings are better than features consisting of high level perceptual abstractions only.

As in [9] and [3], AAREACT coordination layer uses an RL algorithm for achieving behavior coordination, but has a distinct conception. While those architectures select the best behavior to be active, characterizing a competitive behavior coordination, AAREACT selects a set of weights to multiply the behaviors output, determining the influence of each of them in the final robot actuation. It allows more flexibility in robot controlling when more than one objective are concurrently aimed at. For example, in the cases when both tasks, obstacle avoidance and goal reaching, must be executed, the best coordination schema should consider the responses of both, goal pursuing and obstacle avoidance behaviors. Another important difference is the application domain. The problems addressed by [9] and [3] involve a sequence of simple independent tasks, such as go forward until reaching an object, grasp that object, rotate until detecting a light source, etc. Here, the problem addressed is the improvement of navigation performance as a hole, which makes this work unique among the architectures adopting the Module-Based RL approach.

However, in Module-Based RL approach, the need for specifying the feature space and the available behaviors remains. Subtask decomposition can easily be used to define behaviors that correspond to solutions for each subtask. Regarding features, associating each of them with behaviors operation conditions has shown to be a satisfactory approach [9, 16, 3]. Anyway, some automatic approaches have been studied. Terada and colleagues [18] proposed an automatic method for feature definition and discretization. In their work, the appropriate features are selected based on the correlation of the features value and the tactile reinforcement achieved during the execution of a certain task. The optimal features granularity is obtained by using a statistical method. However, this approach may loose interest in practical applications. Besides being developed for the specific case of visual-tactile sensory integration, their method requires a prior training that can be very exhaustive.

8.8 Conclusion

This work presents AAREACT, a hybrid adaptive architecture for mobile robots based on reactive behaviors coordinated in a cooperative fashion. Results obtained with AAREACT show that Reinforcement Learning can be successfully used to learn a policy for behavioral coordination for the task of navigation in a previously unknown environment with obstacle avoidance. Although this work concerns this specific task, other abilities can also be included in the architecture, mainly due to its modular approach. Besides, the case studied is a very common application of mobile robots, projected mainly for navigation purposes.

AAREACT learning methodology presents two important characteristics that deserve to be stressed. First, the knowledge acquired in one environment may be used to let the robot actuate satisfactorily in other ones. The other, as previously expected, is that the robot performance improves by specializing the policy for a specific scenario while it remains actuating in that environment.

However, it is not possible to conclude that the adopted features and weights sets are the best possible. Given the great importance of the problem studied to the robotics community, the investigation on new features and on a less restrictive form of choosing pondering weights may lead to important contributions.

Acknowledgements

Antonio Henrique P. Selvatici acknowledges the support to his PhD course given by the National Council for Scientific and Technological Development — CNPq, through process number 1408682005-4. This work was partially supported by FAPESP, proc. nb. 02/11792-0, and the MultiBot Project, CAPES/GRICES proc. nb. 099/03.

References

1. ActivMedia Robotics, Menlo Park, CA. *Saphira's Manual*, 2001. Version 8.0a.
2. Ronald C. Arkin. *Behavior-Based Robotics.* The MIT Press, Cambridge, MA, 1998.
3. Esther Luna Colombini and Carlos Henrique Ribeiro. An analysis of feature-based and state-based representations for module-based learning in mobile robots. In *Proceedings of the 5th International Conference on Hibrid Intelligent Systems*, pages 163–168, Rio de Janeiro, Brazil, November 2005. IEEE Computer Society.
4. Erann Gat. Integrating planning and reacting in a heterogeneous asynchronous architecture for controlling real-world mobile robots. In *Proceedings of the Tenth National Conference on Artifical Intelligence*, pages 809–815, San Jose, California, 1992.
5. Erann Gat. On three-layer architectures. In D. Kortenkamp, R. Bonasso, and R. Murphy, editors, *Artificial Intelligence and Mobile Robots: Case Studies of Successful Robot Systems*, pages 195–210. MIT Press, 1998.
6. Kousuke Inoue, Jun Ota, Tomohiko Katayama, and Tamio Arai. Acceleration of reinforcement learning by a mobile robot using generalized rules. In *Proceedings of the 2000 IEEE/RSJ International Conference on Intelligent Robots and Systems (IROS'00)*, volume 2, pages 885–890, 2000.
7. Hiroshi Ishiguro, Toshiyuki Kanda, Katumi Kimoto, and Toru Ishida. A robot architecture based on situated modules. In *Proceedings of the 1999 IEEE/RSJ International Conference on Intelligent Robots and Systems (IROS'99)*, volume 3, pages 1617–1624, 1999.
8. Leslie P. Kaelbling, Michael L. Littman, and Andrew Moore. Reinforcement learning: A survey. *Journal of Artificial Intelligence Research*, 4:237–285, 1996.

9. Zsolt Kalmár, Csaba Szepesvári, and András Lörincz. Module-based reinforcement learning: Experiments with a real robot. *Machine Learning*, 31(1–3):55–85, April 1998.
10. Kian Hsiang Low, Wee Kheng Leow, and Marcelo H. Ang Jr. A hybrid mobile robot architecture with integrated planning and control. In *Proceedings of the First International Joint Conference on Autonomous Agents and Multi-Agent System (AAMAS'02)*, pages 219–226, Bologna, Italy, 2002.
11. Maja J. Matarić. Reward functions for accelerated learning. In W. W. Cohen and H. Hirsh, editors, *International Conference on Machine Learning*, pages 181–189. Morgan Kauffman Publishers, Inc., 1994.
12. Robin Murphy. *Introduction to AI Robotics.* The MIT Press, Cambridge, MA, 2000.
13. Ananth Ranganathan and Sven Koenig. A reactive robot architecture with planning on demand. In *Proceedings of the IEEE/RSJ International Conference on Intelligent Robots and Systems (IROS'03)*, volume 2, pages 1462–1468, Las Vegas, California, 2003.
14. Julio Rosenblatt. DAMN: A distributed architecture for mobile navigation. *Journal of Experimental and Theoretical Artificial Intelligence*, 9(2/3):339–360, 1997.
15. Stuart Russell and Peter Norvig. *Artificial Intelligence: A Modern Approach.* Prentice Hall, Upper Saddle River, New Jersey, 2nd edition, 2003.
16. Antonio Henrique Pinto Selvatici. AAREACT: Uma arquitetura comportamental adaptativa para robôs móveis que integra visão, sonares e odometria. Master's thesis, Escola Politécnica da USP, São Paulo, Brazil, February 2005.
17. Richard S. Sutton and Andrew G. Barto. *Reinforcement Learning: An Introduction.* MIT Press, Massachussets, MA, 1998.
18. Kazunori Terada, Takayuki Nakamura, Hideaki Takeda, and Toyoaki Nishida. A congnitive robot architecture based on tactile and visual information. *Advanced Robotics*, 13(8):767–777, 2000.

9

Collaborative Robots for Infrastructure Security Applications

Yi Guo[1], Lynne E. Parker[2], and Raj Madhavan[3]

[1] Department of Electrical and Computer Engineering, Stevens Institute of Technology, Hoboken, NJ 07030, USA, yguo1@stevens.edu, http://www.ece.stevens-tech.edu/~yguo

[2] Department of Computer Science, University of Tennessee, Knoxville, TN 37996, USA, parker@cs.utk.edu, http://www.cs.utk.edu/~parker

[3] Computational Sciences Division, Oak Ridge National Laboratory, Oak Ridge, TN 37831, USA, and Guest Researcher at the Intelligent Systems Division, National Institute of Standards & Technology (NIST), 100 Bureau Drive, Gaithersburg, MD 20899-8230, USA, raj.madhavan@ieee.org

We discuss techniques towards using collaborative robots for infrastructure security applications. A vast number of critical facilities, including power plants, military bases, water plants, air fields, and so forth, must be protected against unauthorized intruders. A team of mobile robots working cooperatively can alleviate human resources and improve effectiveness from human fatigue and boredom. This chapter addresses this scenario by first presenting distributed sensing algorithms for robot localization and 3D map building. We then describe a multi-robot motion planning algorithm according to a patrolling and threat response scenario. Neural network based methods are used for planning a complete coverage patrolling path. A block diagram of the system integration of sensing and planning is presented towards a successful proof of principle demonstration. Previous approaches to similar scenarios have been greatly limited by their reliance on global positioning systems, the need for the manual construction of facility maps, and the need for humans to plan and specify the individual robot paths for the mission. The proposed approaches overcome these limits and enable the systems to be deployed autonomously without modifications to the operating environment.

9.1 Introduction

The events of September 11, 2001 on United States soil have greatly increased the need to safeguard the country's infrastructure. A vast number of critical facilities need to be guarded from unauthorized entry. Unfortunately, the

Yi Guo et al.: *Collaborative Robots for Infrastructure Security Applications*, Studies in Computational Intelligence (SCI) **50**, 185–200 (2007)
www.springerlink.com

number of security officials required to protect these facilities far exceeds their availability. Due to the enormity of this task, it seems unlikely that sufficient human resources can be committed to this infrastructure protection. An alternative approach is to allow technology to assist in this protection, through the use of multiple mobile robots capable of collaborating to guard the grounds of these important facilities from intrusion. Multi-robot systems can thus alleviate the onerous tasks faced by law enforcement officials and army personnel in surveillance, infrastructure security and monitoring of sensitive national security sites (e.g. nuclear facilities, power and chemical plants), building and parking lot security, warehouse guard duty, monitoring restricted access areas in airports and in a variety of military missions.

The Mobile Detection Assessment and Response System (MDARS) described in [7] was developed to provide an automated intrusion detection and inventory assessment capability for use in Department of Defense (DoD) warehouses and storage sites in United States. In this research, the operating area is previously mapped and the positions of the principal features of navigational interest are known in advance. The major sensory characteristics of these features are assumed to be known. By monitoring the variable features of the environment, an intrusion threat is detected. The system adopts random patrols in the secured area.

Another significant work in the area is described in [2], which details a robotic perimeter detection system where a cooperating team of six sentry vehicles are employed to monitor alarms. Formation of vehicles is achieved by teleoperation, while navigation of vehicles to a specified location is achieved by having robots use DGPS (Differential Global Positioning System) to follow specific paths defined by the human. These vehicles have also been used to remotely surround a specified facility. Mission planning is again achieved with the aid of an operator. An operator in the base station uses a graphic interface to determine paths for individual robots and develops a plan outlining obstacles and goal perimeters. The robots then execute this plan by following their designated paths. There are two important disadvantages of this approach from the perspectives of sensing and planning:

- The success of the mission is entirely dependent on positioning information provided by DGPS. *Multipathing*[4] errors make it extremely difficult in many environments to obtain position estimates based on DGPS alone. Thus it becomes necessary to develop a scheme in which observations from relative and absolute sensors can be *fused* to continually deliver reliable and consistent position information for satisfying dynamic mission and motion planning requirements.
- Mission and path planning are fully dependent on the human operator and the system is incapable of dealing with dynamic situation changes that

[4] Multipathing refers to the situation where the signals detected by the DGPS receiver have been reflected off surfaces prior to detection instead of following the straight line path between the satellite and the receiver.

require quick responses and mission and/or path replanning. Even if we did want to use a human operator to specify robot patrol routes, it will be quite difficult for a human to subdivide the patrol region amongst the robots to maximize efficiency. In this case, techniques are needed for dynamic multi-robot motion planning as an aid to the human for determining the best routes to provide to the robots.

There is little work on general frameworks or techniques developed for infrastructure security applications using collaborative robot teams. In this chapter, we formulate the research problems from an infrastructure security scenario, and report on new developments in distributed sensing and motion planning towards such important applications.

9.2 Infrastructure Security Scenario and Research Problems

We envision our new research advances to be used in an infrastructure security scenario such as the following. In an outdoor environment, robot teams are first sent out in a training phase to use our distributed sensing and positioning approach to build 3D digital elevation and obstacle maps of the area to be secured. Once the terrain is learned, the robots will be put into operation and each will operate in one of two modes:

1. a nominal patrol mode,
2. a threat response mode.

In general, robots will operate most often in the nominal patrol mode. In this mode, robots will use our new dynamic multi-robot motion planning algorithms to select efficient multi-robot patrol patterns. Each robot will then patrol its selected region. For efficient patrolling of each area, patrol paths need to be planned according to terrain features and local maps to achieve efficiency. While patrolling, the robots monitor their individual coverage areas for intrusion and also update their local terrain maps to account for environmental changes (e.g., changes in positions of authorized equipment, vehicles, etc.). If an intrusion is detected, some of the robots enter the threat response mode, as pre-defined by the rules of engagement set forth at the beginning of the team deployment. One example of a threat response would be for the detecting robot to send an alert to the human monitor (who is at a remote location), and then for a few robots to surround the threat and return video from multiple perspectives. To successfully respond to the threat, the robots need to dynamically plan paths to the threat location so that they reach the threat area in the shortest possible time. The remaining robots must subsequently replan their patrol paths to compensate for the robots that have entered the threat response mode.

The development of multi-robot teams for use in real world security applications in unstructured outdoor environments presents several challenging issues, which include the following three key research problems:

1. distributed sensing for robot localization and 3D map building,
2. dynamic multi-robot motion planning,
3. integration of approaches to generate a proof of principle demonstration in a relevant infrastructure security environment.

Algorithms need to be developed for collaborative robots to operate in a reliable and robust manner and to be capable of operating in unstructured and dynamic environments with minimal modifications to the operating domain. In the following sections, we address each of these three research problems, presenting approaches and algorithms.

9.3 Multi-Robot Positioning and Mapping using Distributed Sensing

To accomplish missions in infrastructure security applications, multi-robot teams should be able to both autonomously position themselves and construct 3D elevation maps for efficient path planning when traversing on rugged uneven terrain. The objective is to design distributed sensing techniques and to develop schemes that ensure efficient utilization of sensor data obtained from sensors situated across the team members for multi-robot positioning and 3D elevation mapping. The sensors that are considered are: DGPS, wheel encoders, scanning laser rangefinders, inclinometers, compass and pan-tilt-zoom cameras.

9.3.1 Heterogeneous Distributed Multi-Robot Localization

To achieve real-time multi-robot cooperative positioning and mapping competency in a reliable and robust fashion, the sensing and the ensuing data fusion processes are of utmost importance. Thus, careful attention needs to be devoted to the manner in which the sensory information is integrated and interpreted. To satisfy this requirement, we propose a distributed multi-robot Extended Kalman Filter (EKF) estimation-theoretic scheme that enables efficient data fusion of sensor measurements from dead-reckoning and absolute sensors to *continually* deliver reliable and consistent pose (position and orientation) estimates. The robots collect sensor data regarding their own motion and share this information with the rest of the team during the EKF update cycles. The EKF processes the individual positioning information available from all the members of the team and produces a pose estimate for every one of them. Once pose estimates are available, a 3D map of the terrain can be generated by combining vision-based depth estimates with an elevation profile. The elevation profile may be obtained by fusing vertical displacements from DGPS with those computed from inclinometer pitch angles. The proposed scheme has several advantages. The uncertainty associated with measurements from different sensors is explicitly taken into account by using appropriate sensor models and validation procedures. It also becomes possible

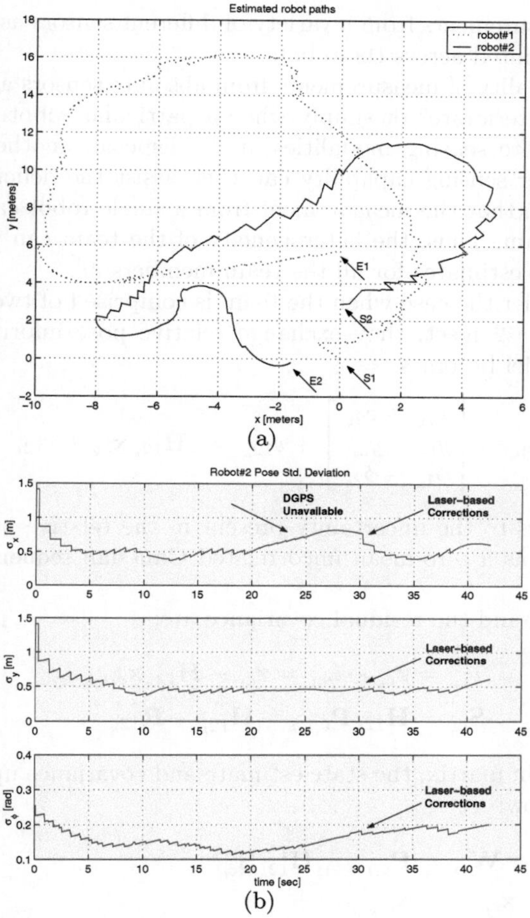

Fig. 9.1. The robots perform laser-based cooperative localization when DGPS becomes unavailable or when there are not enough satellites in view. EKF estimated robot paths are shown in (a). The solid line denotes the estimated path of robot #2 and the dotted line that of robot #1. (S1,E1) and (S2,E2) denote the start and end positions for robots #1 and #2, respectively. The standard deviations of the pose of robot #2 during laser-based cooperative localization are shown in (b). The external corrections offered by the laser-based localization scheme are marked by arrows.

to combine measurements from a variety of different sensors as the estimation process is distributed across the robots.

When the quality of measurements from absolute sensors aboard the individual robots deteriorate[5] or simply when a particular robot of a team does not have adequate sensing modalities at its disposal, another robot in the team with better sensing capability can then assist the deficient member(s) of the team such that the measurement from a single robot can be beneficial to the whole team. Thus, the heterogeneity of the team can be exploited to provide position estimates for all the team members [9].

Let us consider the case when the team is comprised of two robots. When robots #1 and #2 meet, they exchange relative pose information and the observation model becomes:

$$\mathbf{z}_{c_k} = \begin{bmatrix} x_{1_k} - x_{2_k} \\ y_{1_k} - y_{2_k} \\ \phi_{1_k} - \phi_{2_k} \end{bmatrix} + v_{12_k} = \mathbf{H}_{12_k}\mathbf{x}_{c_k} + v_{12_k} \tag{9.1}$$

where v_{12_k} refers to the uncertainty present in the relative pose observation and is modeled as a zero-mean uncorrelated Gaussian sequence with covariance $\mathbf{R}_{12_k}$.

The residual and the residual covariance are:

$$\begin{aligned} \nu_{c_k} &= \mathbf{z}_{c_k} - \hat{\mathbf{z}}_{c_k} = \mathbf{z}_{c_k} - \mathbf{H}_{12_k}\mathbf{x}_{c_{(k|k-1)}} \\ \mathbf{S}_{c_k} &= \mathbf{H}_{12_k}\mathbf{P}_{c_{(k|k-1)}}\mathbf{H}_{12_k}^T + \mathbf{R}_{12_k} \end{aligned}$$

The Kalman gain matrix, the state estimate and covariance updates (centralized) are as below:

$$\begin{aligned} \mathbf{W}_{c_k} &= \mathbf{P}_{c_{(k|k-1)}}\mathbf{H}_{12_k}^T\mathbf{S}_{c_k}^{-1} \\ \mathbf{x}_{c_{(k|k)}} &= \mathbf{x}_{c_{(k|k-1)}} + \\ &\quad \mathbf{W}_{c_k}\left[\mathbf{z}_{c_k} - \left(\mathbf{x}_{1_{(k|k-1)}} - \mathbf{x}_{2_{(k|k-1)}}\right)\right] \\ \mathbf{P}_{c_{(k|k)}} &= \mathbf{P}_{c_{(k|k-1)}} - \mathbf{W}_{c_k}\mathbf{S}_{c_k}\mathbf{W}_{c_k}^T \end{aligned}$$

where $\mathbf{x}_{c_{(k|k-1)}}$ and $\mathbf{P}_{c_{(k|k-1)}}$ are the state and covariance predictions, respectively.

Suppose that robot #2 has a scanning laser rangefinder and also that the number of satellites in view from the current position of this robot indicates that DGPS is unavailable. (In the field trial, this corresponded to the robot going under a tree.) Given the pose of robot #1 whose on-board sensors indicate a high level of confidence in their measurements, relative pose between robots #2 and #1 is determined as follows:

[5] The error in the DGPS positions can be obtained as a function of the number of satellites acquired and this error can then be used as an indicator of the deterioration of the quality of the sensor. As an alternative, the so-called *dilution of precision* measure associated with the GPS can be used for the same purpose.

• Robot #2 identifies robot #1 and acquires a range and bearing laser scan.
• Robot #1 communicates its pose to robot #2.
• After necessary preprocessing to discard readings that are greater than a predefined threshold, the range and bearing to the minima identified in the laser profile of robot #1 are determined.
• From the range and bearing pertaining to the minima, the pose of robot #2 is then inferred.
• Since robot #1 makes its pose available to robot #2, relative pose information is obtained by comparing the two poses and is now available for use in Equation (9.1).

Within the EKF framework, state prediction takes place on individual robots in a decentralized and distributed fashion. By exchanging relative pose information, the states of the robots are then updated in a centralized fashion. The results for the laser-based cooperative localization are shown in Figures 9.1(a) and (b). Figure 9.1(a) shows the estimated paths of robots #1 and #2. The pose standard deviations of robot#2 in Figure 9.1(b) demonstrate the utility of the relative pose information in accomplishing cooperative localization. At $time = 21$ seconds, DGPS becomes unavailable as indicated by the rise in the x standard deviation. It can be seen that as a result of the laser-based relative position information, there is a sharp decrease in the position standard deviations of robot #2 (marked by arrows). As the motion of the robot is primarily in the x direction when the corrections are provided, the resulting decrease in the x standard deviation is noticeable compared to those in y and ϕ.

9.3.2 Terrain Mapping

Incremental terrain mapping takes place via four main processes:

- An incremental dense depth-from-camera-motion algorithm is used to obtain the depth to various features in the environment. The relative pose of the vehicles at these locations as well as depth covariances are associated with particular depth information. These covariances are used to determine regions which contain features of interest that should be indicated on the map.
- An elevation gradient of the terrain is determined by fusing GPS altitude information with vertical displacements obtained from inclinometer pitch angles. • The depth and elevation information are then registered with their associated covariances.
- The terrain map is updated to incorporate the registered values at their proper coordinates. The covariances associated with each measurement provide the confidence the algorithm has in that measurement. In the case of overlapping areas, this confidence determines whether or not the map is updated.

An overall schematic diagram of the algorithm is given in [3]. Both the elevation profile for the motion segments and the feature locations are mapped, as shown in the partially updated terrain map (Figure 9.2). This Figure shows the elevation profile across the area traversed by each robot (in the locally fixed coordinate frame centered at the DGPS base station location) and prominent features within the robot's field of view during the motion segment are marked on the map.

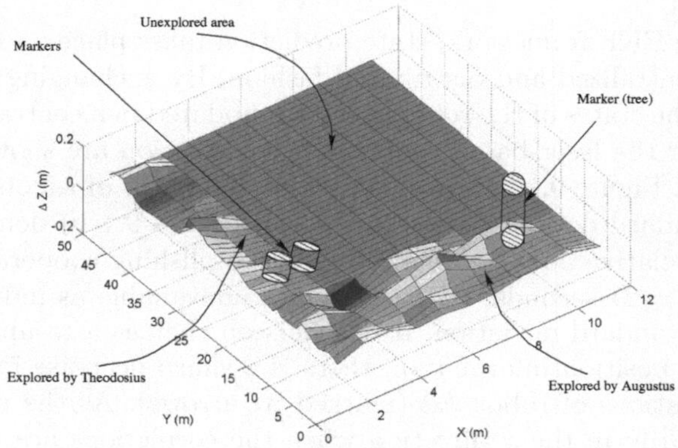

Fig. 9.2. Partially updated terrain map.

9.4 Dynamic Multi-Robot Motion Planning

According to the scenario presented in Section 9.2, the overall patrolling and threat response behavior can be divided into the following design modules:

1. Partition the patrolling region according to the number of robots;
2. Distribute robots from their initial positions to their sub-regions for patrolling;
3. Each robot patrols its sub-region continuously;
4. If a threat is detected by at least one robot during the patrol, a threat alert signal and the threat location are broadcast among robots. A subset of robots move from their current position to the threat position, and the rest of the team repeats steps 1 to 3 to provide continuous patrolling.

In the following, we describe autonomous region partitioning and motion planning in each of the above design modules.

9.4.1 Area Partition

To achieve effective patrolling by a multi-robot team, the first task is to partition an area into sub-areas so that a utility function of the group is minimized. Mathematically, we formulate the problem as follows:

Consider a metric space Q, and n robots with their positions at $\{p_1, p_2, \ldots, p_n\}$. For any point $q \in Q$, assume there is a cost function $f(q, p_i)$, $i \in [1, 2, \ldots, n]$ associated with it. If

$$f(q, p_i) < f(q, p_j),\ i, j \in [1, 2, \ldots, n],\ i \neq j \tag{9.2}$$

we define the decentralized cost function:

$$f_i(q, p_i) = f(q, p_i).$$

A group utility function is defined by

$$U(p_1, p_2, \ldots, p_n) = \sum_{i=1}^{n} \int_Q f_i(q, p_i) dq \tag{9.3}$$

The objective is to find solution $(p_1, p_2, \ldots, p_n)$ so that the group utility function U is minimized.

We know that the set $(p_1, p_2, \ldots, p_n)$ satisfying

$$\frac{\partial U(p_1, p_2, \ldots, p_n)}{\partial p_i} = 0 \tag{9.4}$$

is the solution for $\min U(\cdot)$.

In a special case when Q is a finite dimensional Euclidean space, and the cost function $f(q, p_i)$ is chosen to be the distance between q and p_i, *i.e.*, $f(q, p_i) = dist(q, p_i)$, the set of points satisfying (9.2) compose Voronoi region $V_i = V_i(p_i)$. The set of regions $\{V_1, V_2, \ldots, V_n\}$ is called the Voronoi diagram for the generators $\{p_1, p_2, \ldots, p_n\}$. In this case, the solution to (9.4) is the mass centroid of V_i. Centroidal Voronoi tessellations [1] provides solution methods to find the mass centroids.

Applying the above theoretic results to the area partition, we generate the Voronoi diagram and the mass centroids p_i, $i \in [1, 2, \ldots, n]$, which are the closest point to every point in the Voronoi region V_i.

9.4.2 Initial Distribution

After the set of points $\{p_1, p_2, \ldots, p_n\}$ are generated, we need to move the ith robot from its initial position to p_i, so that the robot can patrol the Voronoi region V_i. The motion planning problem for this sub-task is defined as follows:

Find feasible trajectories for the robot, enrouting from its initial position to its goal position p_i, without collisions with static and dynamic obstacles in the environment while satisfying nonholonomic kinematics constraints.

Solution methods can be found in [6, 12].

9.4.3 Complete Coverage Patrolling

In this sub-task, each robot needs to plan its patrolling path in its own Voronoi region V_i. Multiple paths can be planned based on different criterion, for example, complete coverage of the area as high a frequency as possible, maximize area covered in unit time, minimize repeat coverage, *etc.* For complete coverage, the robot patrols the region so that every point in the region is covered within the robot sensor range at least once over a time period.

We propose a path planning algorithm for complete region coverage. It first packs the bounded region with disks of radius R_c. It was shown in [5] that the disk placement pattern in Figure 9.3 has a minimum number of disks to cover a rectangle. Since the radius of the disk is the same as the coverage range of robot's sensors, complete coverage with minimum repeated coverage can be achieved by visiting every center of the disks. Complete coverage path planning is then to find the sequence to visit the centers. A path planning algorithm was proposed to find a complete coverage path in [5]. However, the algorithm works only in environments without obstacles. Neural network models were used for robot motion planning in dynamic environments in [10] since the dynamically varying environment can be represented by the dynamic neural activity generated by the model. In [13], a neural network approach was developed for complete coverage path planning in a nonstationary environment. We modify the algorithm corresponding to our data structure and generate collision-free complete coverage paths.

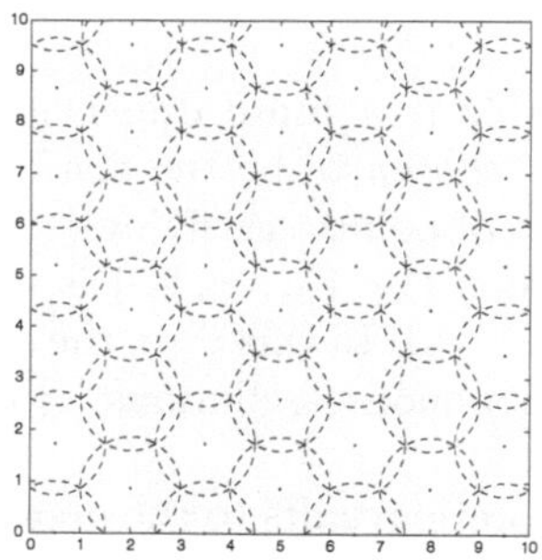

Fig. 9.3. Covering a rectangle using a minimum number of disks

The basic idea of the neural network approach is to generate a dynamic landscape for the neural activities, such that through neural activity propagation, the uncleaned areas[6] globally attract the robot in the entire state space, and the obstacles locally repel the robot to avoid collisions. The dynamics of each neuron in the topologically organized neural network is characterized by a shunting equation derived from Hodgkin and Huxley's membrane equation

[6] The uncleaned areas are defined as the uncovered disks, and the obstacle areas are defined as the disks that are occupied by obstacles.

[8]. The robot path is autonomously generated from the activity landscape of the neural network and the previous location. The neural network model is expressed topologically in a discretized workspace. The location of the neuron in the state space of the neural network uniquely represents an area. In the proposed model, the excitatory input results from the unclean areas and the lateral neural connections, whereas the inhibitory input results from the obstacles only. The dynamics of the neuron in the neural network is characterized by (9.5):

$$\begin{aligned} \dot{x}_i = & -Ax_i + (B - x_i)\left([I_i]^+ + \textstyle\sum_{j=1}^{k} w_{ij}[x_j]^+\right) \\ & -(D + x_i)[I_i]^- \end{aligned} \qquad (9.5)$$

where k is the number of neural connections of the ith neuron to its neighboring neurons within the receptive field. Six neighbors of a point (neuron) are shown in Figure 9.4.

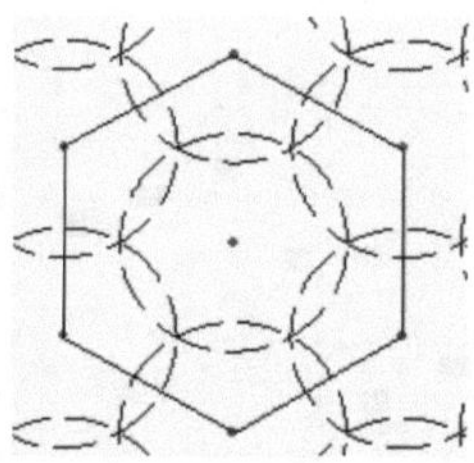

Fig. 9.4. Six neighbors of a neuron

The external input I_i to the ith neuron is defined as in (9.6).

$$I_i = \begin{cases} E & \text{if it is an unclean area} \\ -E & \text{if it is an obstacle area} \\ 0 & \text{otherwise} \end{cases} \qquad (9.6)$$

where $E >> B$ is a very large positive constant. The terms $[I_i]^+ + \sum_{j=1}^{k} w_{ij}[x_j]^+$ and $[I_i]^-$ are the excitatory and inhibitory inputs respectively. Function $[a]^+$ is a linear threshold function defined as $[a]^+ = \max\{a, 0\}$, and $[a]^- = \max\{-a, 0\}$. The connection weight w_{ij} between the ith and the jth neurons is 1 if they are neighbors or 0 if they are not neighbors.

To make the path having less navigation turns, for a current robot location p_c, we select the next point p_n within the uncleaned neighbors according to (9.7):

$$x_n = \max\{x_j + (1 - \frac{\Delta\theta_j}{\pi}),\ j = 1, 2, \ldots, k\} \qquad (9.7)$$

where k is the number of neighboring neurons, x_j is the neuron activity of the jth neuron, $\Delta\theta_j$ is the absolute angle change between the current and next moving directions, i.e., $\Delta\theta_j = 0$ if going straight, and $\Delta\theta_j = \pi$ if going backward. After the robot reaches its next position, the next position becomes a new current position. Because of the excitatory neural connections in (9.5), the neural activity propagates to the entire state space so that the complete coverage is achieved.

A complete coverage path in an environment with stationary obstacles is shown in Figure 9.5. It is shown that the path completely covers the bounded region without covering a point twice. Note that in a trap situation, that is, there are no uncleaned neighbors, the neighbor's neighbors become the neighbors of the neuron, so that uncleaned area can be continuously searched. This can be seen from the right bottom and top middle parts of the figure. The algorithm works for moving obstacles if the speed of the obstacles are known, since the landscape activities of the environment are updated dynamically in the algorithm.

Cooperative patrolling paths of four robots are shown in Figure 9.6.

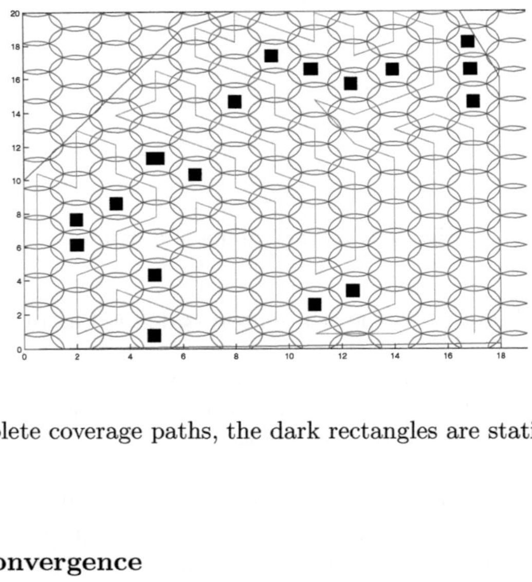

Fig. 9.5. Complete coverage paths, the dark rectangles are stationary obstacles

9.4.4 Point Convergence

In this module, a subset of robots move to the threat location from their current positions. The motion planning problem for each robot becomes: *given a start and a goal, generate a feasible trajectory without collisions.* The same solution methods as described in subsection 9.4.2 can be applied.

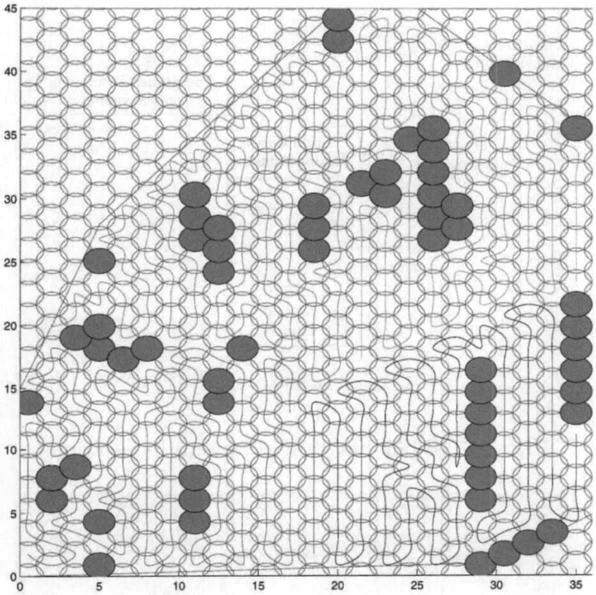

Fig. 9.6. Cooperative coverage trajectories: each continuous curve represents one robot's trajectory, and the solid circles denote areas occupied by stationary obstacles.

9.5 System Integration Towards Proof of Principle Demonstration

For a successful proof of principle demonstration in a relevant infrastructure security environment, functionalities of distributed sensing developed in Section 9.3 should be integrated with the dynamic motion planning capabilities in Section 9.4 to realize the cooperative team objectives. Figure 9.7 illustrates the block diagram of system integration. Note that in the figure, techniques regarding threat detection (in the gray box) are not discussed in this chapter.

We have partially implemented the algorithms proposed on a group of ATRV-mini robots. The experimental setup is shown in Figure 9.8. It consists of a wireless mini-LAN, a Local Area DGPS (LADGPS), a software platform (*Mobility* from RWI) and codes developed in-house under Linux to read and log the data for the sensors on each robot. The wireless LAN is set up outdoors between an Operator Console Unit (OCU) and the robots. The OCU consists of a rugged notebook equipped with a BreezeCOM access point and antenna. Each robot has a BreezeCOM station adaptor and an antenna. The LADGPS is formed by the base station/antenna hardware connected to the OCU and

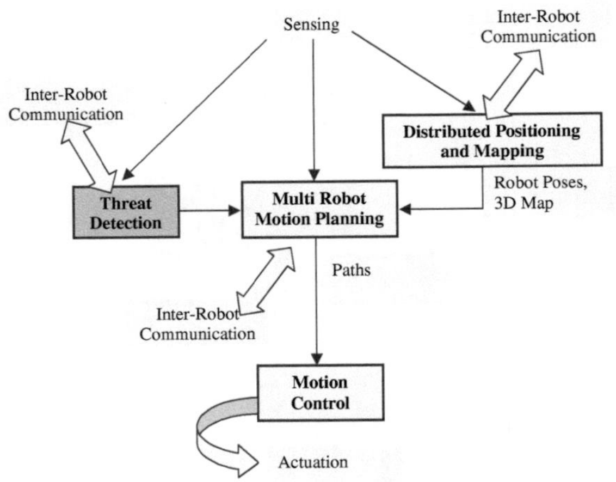

Fig. 9.7. Block diagram of system integration.

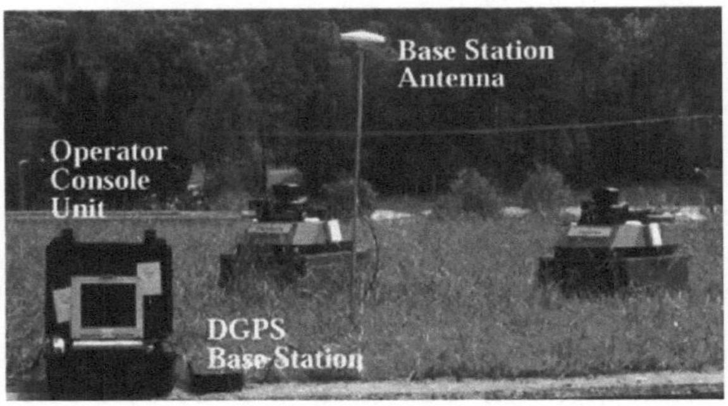

Fig. 9.8. Experimental setup in an outdoor environment

remote stations/antennas directly mounted on each robot. Each robot's station receives differential corrections from the base station such that LADGPS accuracy of up to 10 centimeters is obtainable. Some experimental results can be found in our previous publications [3, 4, 9, 11].

9.6 Conclusions

Recent terrorist events on United States soil have dramatically increased the need for protection of our nation's infrastructure. Rather than stretch already-thin human resources to guard facilities against low-probability intrusions, technological solutions are preferred. We propose to address this problem by using teams of intelligent robots for infrastructure security applications.

We first formulate the research problems from an infrastructure scenario, and then propose new algorithms in distributed sensing and multi-robot motion planning to achieve the autonomous patrolling and threat response tasks. Finally, the system integration of sensing and planning are presented towards a successful proof of principle demonstration. The developed collaborative sensing and motion control strategies enable a robot team to position themselves and move appropriately in a previously unknown environment to enable intrusion detection. To briefly summarize the advances of our approach over existing approaches for infrastructure security in outdoor environments, we compare different approaches in Table 9.1. Future work includes extensive real robot experiments using the experimental setup in Section 9.5.

Table 9.1. Comparison of approaches

	Proposed Approach	Approach in [2]	Approach in [7]
Autonomous detection of threats	X	X	X
Mobility for rapid response	X	X	X
No infrastructure modifications required	X	X	
No dependency on absolute positioning (DGPS)	X		
Autonomous path planning	X		
No a priori map needed	X		

References

1. Q. Du, V. Faber, and M. Gunzburger. Centroidal Voronoi tessellations, theory, algorithms and applications. *SIAM Review*, 41:637–676, 1999.
2. J. T. Feddema, C. Lewis, and P. Klarer. Control of multiple robotic sentry vehicles. In *Proceedings of SPIE Unmanned Ground Vehicle Technology, SPIE*, volume 3693, pp. 212–223, 1999.
3. K. Fregene, R. Madhavan, and L.E. Parker. Incremental Multiagent Robotic Mapping of Outdoor Terrains. In *Proceedings of the IEEE International Conference on Robotics and Automation*, pp. 1339–1346, May 2002.
4. Y. Guo and L. E. Parker. A distributed and optimal motion planning approach for multiple mobile robots. In *Proceedings of IEEE International Conference on Robotics and Automation*, pp. 2612–2619, May 2002.
5. Y. Guo and Z. Qu. Coverage control for a mobile robot patrolling a dynamic and uncertain environment. In *Proceedings of World Congress on Intelligent Control and Automation*, pp. 4899–4903, China, June 2004.
6. Y. Guo, Z. Qu, and J. Wang. A new performance-based motion planner for nonholonomic mobile robots. In *Proceedings of the 3rd Performance Metrics for Intelligent Systems Workshop*, NIST, Gaithersburg, MD, Sept. 2003.
7. T. Heath-Pastore, H.R. Everett, and K. Bonner. Mobile Robots for Outdoor Security Applications. In *Proceedings of the American Nuclear Society 8th International Topical Meeting on Robotics and Remote Systems*, April 1999.

8. A. L. Hodgkin and A. F. Huxley. A quantitative description of membrane current and its application to conduction and excitation in nerve. *J. Physiol. Lond.*, 117:500–544, 1952.
9. R. Madhavan, K. Fregene, and L.E. Parker. Distributed Cooperative Outdoor Multirobot Localization and Mapping. In *Autonomous Robots (Special Issue on Analysis and Experiments in Distributed Multi-Robot Systems)*, Vol. 17, Issue 1, pp. 2339, 2004.
10. F. Muniz, E. Zalama, P. Gaudiano, and J. Lopez-Coronado. Neural controller for a mobile robot in a nonstationary environment. In *Proceedings of 2nd IFAC Conference on Intelligent Autonomous Vehicles*, Helsinki, Finland, pp. 279-284, 1995.
11. L. E. Parker, K. Fregene, Y. Guo, and R. Madhavan. Distributed heterogeneous sensing for outdoor multi-robot localization, mapping, and path planning. In A. C. Schultz and L. E. Parker, editors, *Multi-Robot Systems: From Swarms to Intelligent Automata*, pp. 21–30. Kluwer, The Netherlands, 2002.
12. Z. Qu, J. Wang, and C. E. Plaisted. A new analytical solution to mobile robot trajectory generation in the presence of moving obstacles. In *Proceedings of 2003 Florida Conference on Recent Advances in Robotics*, May 2003. Also submitted to IEEE Transactions on Robotics and Automation.
13. S. X. Yang and C. Luo. A neural network approach to complete coverage path planning. *IEEE Transactions on Systems, Man and Cybernetics - Part B: Cybernetics*, 34(1):718–725, 2004.

10

Imitation Learning: An Application in a Micro Robot Soccer Game

Dennis Barrios-Aranibar and Pablo Javier Alsina

Department of Computing Engineering and Automation
Federal University of Rio Grande do Norte
Lagoa Nova 59.072-970 - Natal - RN - Brazil {dennis, pablo}@dca.ufrn.br

In this chapter, we present a robot soccer system that learns by imitation and by experience. At first the robots do not know anything but after observing saved games of other teams, they start improving their knowledge learning new states and actions. This is called learning by imitation. They validate learned actions by testing them several times, which called learning by experience. Repeating this process allows that robots can continuously improve their performance. This strategy emulates the way humans learn.

10.1 Introduction

In robotics several tasks require complex behaviors. Behaviors become more complex when they include interactions between two or more robots. It is the case of a robot soccer game.

Definition 1. *A* **multi-robot system** *is a robotic system including two or more robots interacting between each other and with the environment in a collaborative way.*

An important issue in a multi-robot system is its work strategy, which generally involves individual and collaborative behaviors. There are two ways to implement a work strategy: Statically and dynamically.

A static strategy is defined during the multi-robot system design and doesn't change automatically. A dynamic strategy is improved according to the experience of each component in the system. While a static strategy depends on the designers knowledge and experience about the problem; a dynamic strategy only depends on the way selected to improve it as algorithms, examples of situations, adversary, etc.

D. Barrios-Aranibar and P. J. Alsina: *Imitation Learning: An Application in a Micro Robot Soccer Game*, Studies in Computational Intelligence (SCI) **50**, 201–219 (2007)
www.springerlink.com

In general, dynamic strategies can be divided in two stages, the learning and the operation stage. During the learning stage the system is trained by interacting with a real environment or an environment close to the real one (e.g. a simulated environment) in order to tune all the parameters that define the strategy. In the case of robot soccer teams, a team is trained by playing several times with another team (e.g. a team that plays "well enough"). In robot soccer, previous works that implement dynamic strategies [3, 4, 7, 8, 14, 17, 18, 22] use intelligent systems to accomplish this objective.

A dynamic strategy can be implemented using search algorithms for finding optimal parameters for the strategy or learning algorithms used to learn optimal parameters for the strategy or a optimal strategy. Both implementations can be seen as experience learning because the way they are implemented (e.g. playing several times to tune strategy). The problem in these implementations is that actions and behaviors in robot strategy are restricted to those defined by the system designers. A way to solve this problem is by learning new actions and behaviors from other systems developing the same task. This learning is called imitation learning.

Definition 2. *Imitation Learning is a technique based on the interaction of an agent with others that know something about what the agent is trying to learn. In human agents this interaction is known as social interaction.*

Learning by imitation is considered a method to acquire complex behaviors and as a way to provide seeds for further learning [9, 11, 15]. This type of learning was applied to several problems like humanoid robot learning [10, 16] and air hockey and marble maze games [2]. In robot soccer, learning by imitation was implemented using a Hidden Markov Model [12, 13] and by teaching the robots with a Q-Learning algorithm [6], but not by observing other teams playing soccer for recognizing complex behaviors.

In this chapter, a robot soccer system that learns by imitation and by experience is presented (figure 10.1). Initially, robots do not know anything, then robots improve their knowledge learning new states and actions observing saved games of other teams (learning by imitation). After this, they validate learned actions by testing them several times (learning by experience). This process is repeated exhaustively, thus robots can continuously improve their performance. This combination tries to emulate the way humans learn.

To understand this approach of learning by imitation, the next concepts were defined:

Definition 3. *A* **role** *is defined as the function a robot implements during it's useful life in a problem or in part of it. This can be part of the functions in a robotic team or part of the applications of a particular robot. A role is independent of the robot and the form it is executed depends on the characteristic of the robot implementing it. Each role is composed of a set of behaviors.*

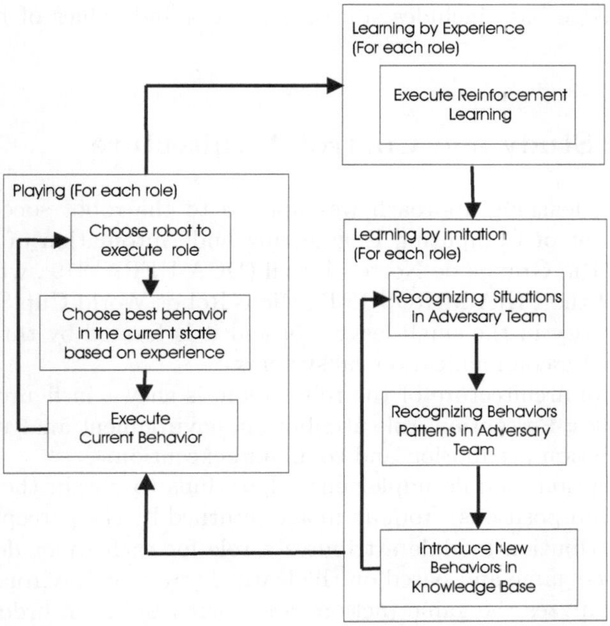

Fig. 10.1. Learning by Imitation and Experience Applied to the Robot Soccer Problem

Definition 4. *A* **situation** *is an observer description of a basic interaction between a robot and an element in the environment, including other robots in the team.*

Definition 5. *An* **action** *is a basic interaction between a robot (actor) and an element in the environment, including other robots in the team. This interaction is intentionally caused by the actor. While for the actor this interaction is an action, for observers it is only a situation because they are not sure if the actor really wants to do that.*

Definition 6. *A* **behavior** *is a sequence of consecutive actions. In this approach, a behavior can be composed of one, two, three or four actions.*

The case study presented in this chapter is the robot soccer standard problem. According to the proposed approach, in the present case study, learning is implemented separately for each role (section 10.2). Situations of a robot are recognized, in a recorded game (section 10.3). Behaviors patterns are founded from groups of consecutive situations executed several times (section 10.4). Finally these behaviors are introduced in the knowledge base. Knowledge base has the form of the environment model in a reinforcement learning problem because learning by experience was implemented using reinforcement learning.

Then, knowledge base includes states, behaviors and values of behaviors at each state.

10.2 Case Study and Control Architecture

The proposed learning approach was applied to the robot soccer team of the Department of Computing Engineering and Automation of the Federal University of Rio Grande do Norte - Brazil (DCA-UFRN) [19], which satisfies restrictions of the FIRA MIROSOFT (Micro Robot World Cup Soccer Tournament) category in the small league [5] and it is formed by three robots, a perception and a communication subsystems.

The control architecture of the robot team is shown in figure 10.2 [1]. It has six modules: Perception, role attribution, environment analysis, behavior attribution, action attribution and command execution.

The perception module implements algorithms to obtain the game state (robots and ball positions) from an image returned by the perception subsystem. Role attribution module attributes a role for each robot depending on the state of the game and based on the learned strategy. Environment analysis module analyzes the game (actions and their results) in order to learn a better strategy. Behavior attribution module attributes a behavior to a robot based on the game state, attributed role, result of executed actions and learned strategy. Action attribution module executes an action based on the game state, attributed role, result of executed actions, learned strategy and attributed behavior. Finally, the command execution module translates a specific action into a sequence of commands (e.g. reference motor voltage) and executes them sending these signals by the communication subsystem.

In this architecture, modules are grouped according to the element (central agent, role or robot) they belong to. Role attribution, environment analysis and Perception modules are part of a central processing. This occurs because in this application environment is perceived using a single central camera connected to a computer (in other applications, perception module can be part of the robot processing). A central agent decides which robot will execute each role, based on robots positions. Finally, this central agent analyzes actions and their results for implement learning by experience. Also it analyzes opponent movements in order to implement imitation learning.

Behavior and action attribution are part of a role processing. Thus, each role has its own states, behaviors and actions and its own strategy to attribute behaviors to the robot executing the role at each state. It occurs because each role implements its own reinforcement learning process. Finally, command execution is part of the robot processing because each robot translate actions into commands depending on its physic characteristics.

In the environment analysis module, both an analysis of the team performance and other team behaviors in recorded games are made. Analysis of team behavior is used to improve the strategy parameters, and, analysis of

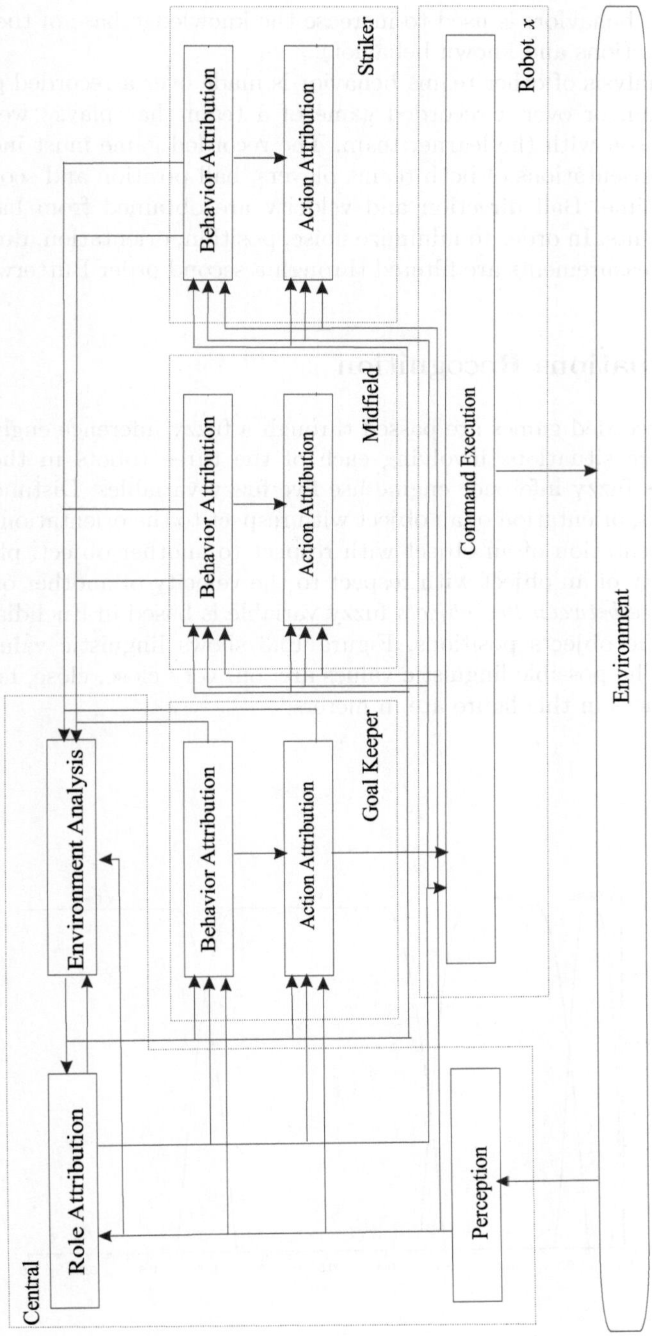

Fig. 10.2. Control Architecture of a Multi-Robot System Applied to the Robot Soccer Problem

other team behaviors is used to increase the knowledge base of the team (set of states, actions and known behavior).

The analysis of other teams behavior is made over a recorded game of an expert team, or over a recorded game of a team that plays "well enough" in comparison with the learner team. The recorded game must include positions and orientations of both teams players, ball position and score at every sampling time. Ball direction and velocity are obtained from ball position measurements. In order to minimize noise, position, orientation, direction and velocity measurements are filtered through a second order Butterworth filter.

10.3 Situations Recognition

Filtered recorded games are passed through a fuzzy inference engine in order to recognize situations involving each of the three robots in the analyzed team. This fuzzy inference engine use five fuzzy variables: Distance between two objects, orientation of an object with respect to the orientation of another object, orientation of an object with respect to another object, playing time and velocity of an object with respect to the velocity of another object.

Distance between two objects fuzzy variable is based in Euclidian distance between the objects positions. Figure 10.3 shows linguistic values for this variable. The possible linguistic values are: on, very close, close, far and very far. Distances in this figure are in meters.

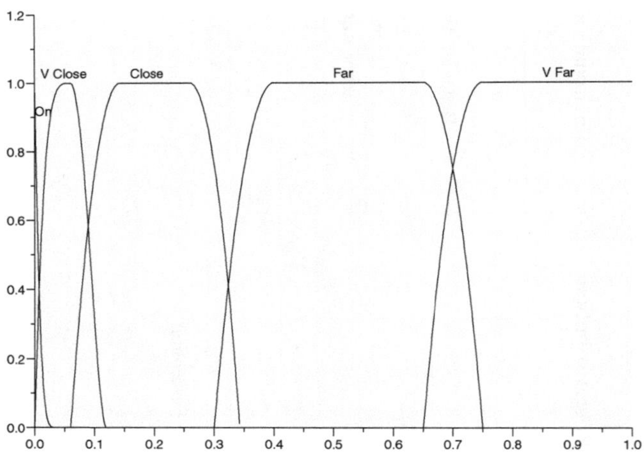

Fig. 10.3. Distance Between two Objects Fuzzy Variable

Orientation of an object with respect to orientation of another object fuzzy variable is based on the absolute value of the difference between the orientation angles of two objects. Figure 10.4 shows linguistic values for this variable. This fuzzy variable has three possible linguistic values: same, almost same and different.

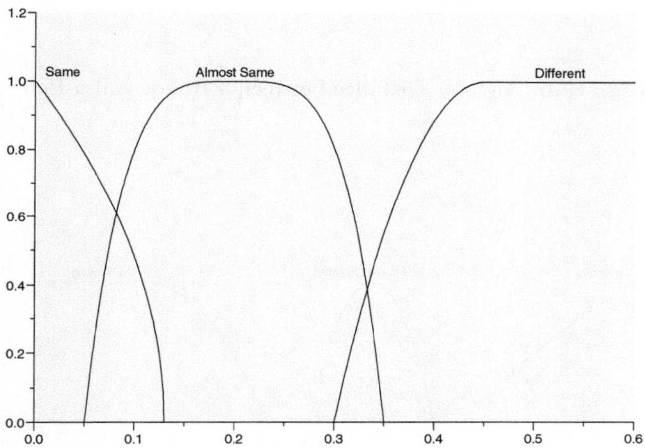

Fig. 10.4. Orientation of an Object with Respect to Orientation of Another Object Fuzzy Variable

Orientation of an object with respect to another object fuzzy variable is based on the angular distance between two objects. It is defined as the difference between the current orientation of the first object and the orientation needed to point to the second object, as shown in figure 10.5. Figure 10.6 shows linguistic values for this variable. This fuzzy variable has three possible linguistic values: oriented, almost oriented and not oriented.

Playing time fuzzy variable is based on the sampled times of a game or a part of it. It was defined using a sample rate of 33 milliseconds. Figure 10.7 shows linguistic values for this variable. This fuzzy variable has five possible linguistic values: very small, small, medium, great and very great. Generally this variable maps situation elapsed time.

Velocity of an object with respect to the velocity of another object fuzzy variable is based on the value of the difference between the velocity of two objects. Figure 10.8 shows linguistic values for this variable. This fuzzy variable has five possible linguistic values: very slow, slow, equal, fast, very fast.

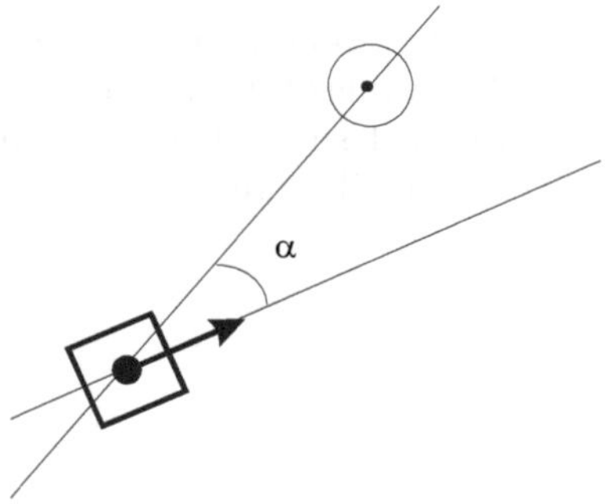

Fig. 10.5. Angular Distance between a Robot and a Ball

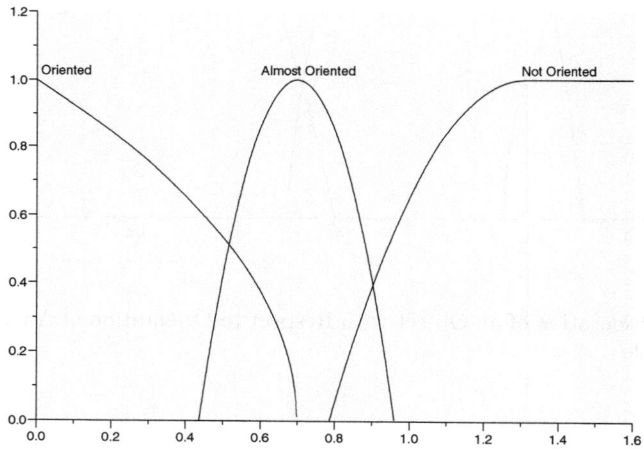

Fig. 10.6. Orientation of an Object with Respect to Another Object Fuzzy Variable

Thirty situations with fuzzy rules were defined. An additional situation called "*Robot moving randomly*" was defined to be used when there is a time interval that does not fulfill restrictions of any situation. Situations are listed in table 10.1.

Three roles were defined for the robots: Goalkeeper, midfield and striker, all of them depend on the robot position at each sampling time. Almost all situations are independent of the robot's role. Situations with codes 162, 163, 173 and 174 depend on the robot role when a situation starts. For that purpose, roles are arranged in a circular list (e.g. Goalkeeper, midfield, striker). Role of analyzed robot, role +1 and role +2 can be goalkeeper, midfield and

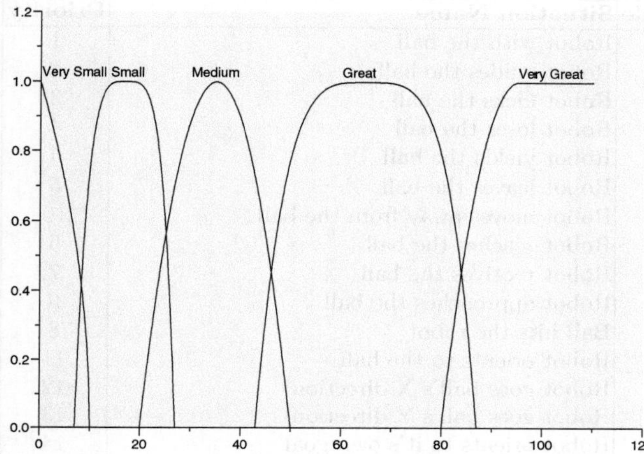

Fig. 10.7. Playing Time Fuzzy Variable

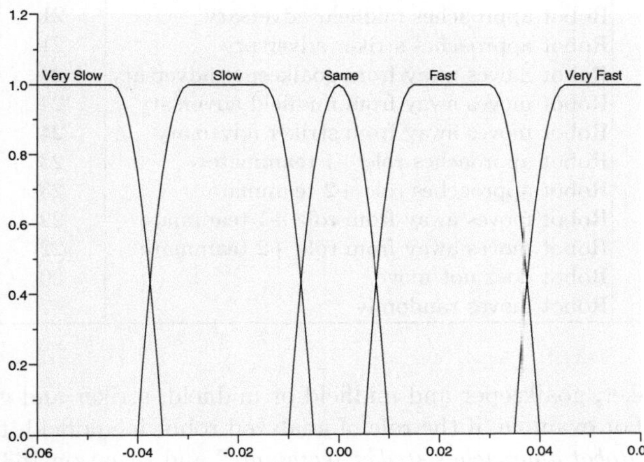

Fig. 10.8. Velocity of an Object with Respect to the Velocity of Another Object Fuzzy Variable

Table 10.1. Situations Defined in the Fuzzy Inference Engine

Code	Situation Name	Priority
001	Robot with the ball	1
011	Robot guides the ball	2
021	Robot kicks the ball	3
022	Robot loses the ball	4
023	Robot yields the ball	4
033	Robot leaves the ball	5
034	Robot moves away from the ball	10
044	Robot reaches the ball	6
045	Robot receives the ball	7
046	Robot approaches the ball	9
047	Ball hits the robot	8
057	Robot orients to the ball	11
067	Robot goes ball's X direction	12
068	Robot goes ball's Y direction	13
078	Robot orients to it's own goal	14
088	Robot approaches it's own goal	15
098	Robot moves away from it's own goal	16
108	Robot orients to adversary's goal	17
118	Robot approaches the adversary's goal	18
128	Robot moves away from adversary's goal	19
138	Robot approaches goalkeeper adversary	21
139	Robot approaches midfield adversary	21
140	Robot approaches striker adversary	21
150	Robot moves away from goalkeeper adversary	24
151	Robot moves away from midfield adversary	24
152	Robot moves away from striker adversary	24
162	Robot approaches role +1 teammate	23
163	Robot approaches role +2 teammate	23
173	Robot moves away from role +1 teammate	22
174	Robot moves away from role +2 teammate	22
184	Robot does not move	20
194	Robot moves randomly	

striker or striker, goalkeeper and midfield or midfield, striker and goalkeeper respectively. For example, if the role of analyzed robot is midfield, then, situation 162 is "*Robot approaches striker teammate*" and situation 163 is "*Robot approaches goalkeeper teammate*", the same occurs with situations 173 and situations 174.

To figure out how fuzzy rules are defined, a sample of "*Robot kicks the ball*" situation is shown in table 10.2. In this fuzzy rule, four fuzzy variables are used: distance between two objects, orientation of an object in relation to the orientation of another object, orientation of an object in relation to another object and velocity of an object in relation to velocity of another object.

Table 10.2. Fuzzy Rule of Situation *Robot Kicks the Ball*

021: Robot kicks the ball
Initial condition : If robot is [(**on** or **very close**) and **oriented**] to the ball, then, situation is set as <*possible to happen*> and the current state is saved as the state in the beginning of the situation. **Final condition :** If situation is set as <*possible to happen*> and ([robot is not (**on** or **very close**) to the ball, and (the ball is going (**fast** or **very fast**) and with the **same** orientation in comparison with the beginning of the situation)] or [the ball is not following the **same** orientation]) and the ball is not **on** the last position and all other robots are not ((**on** or **very close**) and **oriented**) to the ball in any moment from the beginning of the situation to the current time, then robot kicks the ball.

Recognizing situations involve two steps: Verification of possibility and confirmation of occurrence. Verification of possibility involves a fuzzy rule that verifies if initial conditions for a situation to happen are accomplished. For example in order to say that a robot kicks the ball it is important that initially it is near the ball (on or very close) and oriented to it. If this initial condition is fulfilled then the situation is possible to happen.

Confirmation of occurrence is the effective recognition of a situation, which is the accomplishment of final restrictions, expressed as another fuzzy rule, for situations that were first recognized as *possible to happen*.

For example, final conditions of situation 021 involves two options:

- The robot is not on or very close to the ball and the ball is going faster than when the situation begins, but the ball must have the same orientation it had at the beginning of the situation.
- The robot is not near the ball and it changes its direction.

Two restrictions complement this final condition: The ball is moving, which means that its position is changing at every time step. And, the fact that no robot is near and oriented towards the ball during the situation.

When the fuzzy engine finishes recognizing situations, it verifies if there are sampled states that are not found as part of any situation. In the affirmative case, all these subsequent samples are grouped and evaluated with the fuzzy variable playing time. If the time is not **very small** then this group of states are recognized as the situation "*Robot moves randomly*", else the beginning of the situation after this group of states is moved a half of the total amount of sampled states in that group, and the end of the former situation is moved to the other half in order to have both situations in sequence.

After recognizing all situations for each robot of the analyzed team, the codes are passed trough a SOM neural network in order to find patterns of groups of subsequent situations. This analysis is off-line because it is necessary to have a complete game or group of games to find behavior patterns. A game can be analyzed on line if necessary, but only for recognition of situations, because the neural network SOM requires a lot of time to converge in comparison with the sampling time in a robot soccer game. In this case, situation "*Robot moves randomly*" is not utilized.

Each situation has a priority in the inference engine. Situations involving interaction with the ball have greater priority than those involving interaction with other elements of the game, like robots or goals. This is because the objective of this approach is to implement an imitation learning and it is better to learn interaction with the ball than interaction with another element of the game.

10.4 Behaviors Patterns Recognition

The output of the fuzzy inference engine is a list of subsequent situations for each robot in the team. In this stage, the code of each situation will be used. For this purpose a SOM neural network is used. Groups of one, two, three and four situations are used to train this neural network. Codes of situations are used as entries. It means that when it is used to recognize patterns in behaviors composed by n situations the neural network will have n entries.

As shown in table 10.1, codes of similar situations are separated by one unit. For example situation 021 "*Robot kicks the ball*" is similar to the situation 022 "*Robot loses the ball*" because both situations will be translated into action "kick the ball". In both cases the robot wants to kick, but, in the second one it can't complete this action because the presence of an adversary robot. Codes of different situations are separated by ten units. This codification will bring a good performance in the situation groups pattern recognition.

Recognizing patterns in groups of situations is guided by a priority. First are recognized patterns in groups of four situations, next in groups of three, two and one situations respectively.

Groups of four subsequent situations are formed. All possible combinations of subsequent situations are used, that means, each situation can be part of a maximum of 4 groups (e.g. being the first, second, third or fourth in the group).

Formed groups are used to train the SOM neural network. After training process, neurons that were activated by at least ten percent of the groups are selected. Next, groups that activate these selected neurons are selected to be part of the knowledge base. To be part of the knowledge base, these groups must have a value greater than a threshold.

The value of a group of situations is calculated based on its final result: Own goal, adversary goal or game finished without a goal. Each situation preceding an own goal receive a discounted positive value α^t, where α is a value greater or equal than zero and less or equal than one, and t is the number of samples between the beginning of the situation and the final result. Each situation preceding an adversary goal receives a discounted negative value $-\alpha^t$, and each situation preceding the finish of the game without goals receives zero as a value. Value of a group of situations is calculated as the arithmetic mean of the values of each situation in the group.

After recognizing behaviors formed by four situations, groups of three situations are formed with situations not considered as part of behaviors added to the knowledge base. The process explained before is then repeated, but this time neurons that were activated by at least eight percent of the groups are selected. Next, remainder situations are grouped in groups of two and the process is repeated considering neurons that were activated by at least six percent of the groups. Finally the same is made for individual remainder situations with four percent.

Because reinforcement learning will be used to improve strategy (implement learning by experience), each new behavior in the knowledge base receives a value for each state. These behaviors will receive an optimistic value (e.g. the maximum value of the known behaviors in the state). This initial value will assure that each new behavior in each state will be explored and its right value will be found.

10.5 Experimental Results

This approach was implemented and tested in the robot soccer simulator developed by Yamamoto [20], which has the same characteristics of the soccer team of the DCA-UFRN[19].

The Proposed approach was tested by analyzing recorded games of two teams using the static strategy implemented by Yamamoto[21]. It was observed that recognized situations are similar to those recognized by a human observer.

An analysis of the efficiency of the proposed fuzzy inference engine was also performed. In 30 minutes of simulated game, each situation was translated into actions and executed by a robot a random number of times. The other five robots were stopped in random positions. Results of this analysis are shown in table 10.3. Situation 045 was not executed in these 30 minutes because; it involves a complex initial state that was not held. Situations 184 and 194 were not considered to be executed.

As shown in this table, 68.56% of all executed situations were recognized as the effectively executed or as a similar one (efficiency rate). 31.45% of them were recognized as a situation or group of situations without relation to the executed one (error rate).

To be executed, each situation is translated into an action. For example situation "*Robot looses the ball*" is translated into action "kick the ball". This situation will be recognized as "*Robot kicks the ball*". If there are some adversary or teammate robot near and oriented to the ball during its execution then it can be recognized as "*Robot looses the ball*" or "*Robot yields the ball*" respectively.

Table 10.3. Situations Recognition Efficiency Analysis

Code	same(%)	related(%)	different(%)
001	75,00	25,00	0,00
011	24,00	72,00	4,00
021	32,76	36,21	31,03
022	48,53	42,65	8,82
023	7,41	64,81	27,78
033	34,33	44,78	20,90
034	32,26	41,94	25,81
044	21,41	53,43	25,16
046	25,93	69,14	4,94
047	18,31	73,24	8,45
057	45,74	29,79	24,47
067	31,75	52,38	15,87
068	32,31	33,85	33,85
078	10,00	64,44	25,56
088	3,73	74,63	21,64
098	1,49	71,64	26,87
108	17,86	54,76	27,38
118	8,33	76,39	15,28
128	1,54	80,00	18,46
138	2,54	48,31	49,15
139	2,34	39,84	57,81
140	3,57	46,43	50,00
150	3,77	45,28	50,94
151	0,00	53,97	46,03
152	1,89	39,62	58,49
162	0,00	49,60	50,40
163	2,99	47,76	49,25
173	0,00	54,72	45,28
174	2,99	50,75	46,27
ALL	14,37	54,19	31,45

It was observed that each situation can be executed in a different number of ways, depending on the exact position of the robot at the beginning of the situation. It was also observed that the lesser the priority of the situation in the inference engine the greater the error in the results.

This approach was used to narrate a soccer game. Narrating a game includes to say the time each situation starts and ends (minutes: seconds: milliseconds of game). Table 10.4 shows part of an output of this approach used to narrate a soccer game.

An analysis of the number of actions executed in each game using this approach was made. Remember that the learner team start with no knowledge and after each game it learns new actions from its adversary. Table 10.5 shows

Table 10.4. Output of this approach used to narrate a soccer game

...
Blue robot of blue team does not move from (6:35:10) to (6:36:0)
Green robot of blue team does not move from (6:35:175) to (6:37:551)
Cyan robot of blue team does not move from (6:35:208) to (6:37:89)
Blue robot of yellow team goes ball's Y direction from (6:35:307) to (6:37:386)
Green robot of yellow team orients to the ball from (6:35:802) to (6:36:825)
Blue robot of blue team goes ball's Y direction from (6:36:0) to (6:36:660)
Blue robot of blue team approaches to the ball from (6:36:660) to (6:36:858)
Green robot of yellow team does not move from (6:36:825) to (6:50:520)
Blue robot of blue team approaches the adversary's goal from (6:36:858) to (6:37:188)
Cyan robot of blue team reaches the ball from (6:37:89) to (6:37:419)
Blue robot of blue team moves away from the ball from (6:37:188) to (6:37:584)
Cyan robot of yellow team does not move from (6:37:188) to (6:43:821)
Blue robot of yellow team does not move from (6:37:386) to (6:42:666)
Cyan robot of blue team orients to the ball from (6:37:419) to (6:37:518)
Cyan robot of blue team kicks the ball from(6:37:518) to (6:37:584)
Blue team goal!!!! (6:37:584)
Cyan robot of blue team orients to the ball from (6:37:584) to (6:43:425)
Blue robot of blue team does not move from (6:37:584) to (6:43:62)
Blue robot of yellow team goes ball X's direction from(6:42:666) to (6:43:755)
Green robot of blue team does not move from (6:42:930) to (6:43:425)
Blue robot of blue team reaches the ball from (6:43:62) to (6:44:118)
Green robot of blue team goes ball's Y direction from (6:43:425) to (6:43:953)
Cyan robot of blue team does not move from (6:43:425) to (6:43:590)
Cyan robot of blue team goes ball's Y direction from (6:43:590) to (6:44:349)
Blue robot of yellow team moves away from the ball from (6:43:755) to (6:44:679)
Cyan robot of yellow team moves away from the ball from (6:43:821) to (6:44:613)
Green robot of blue team moves away from the ball from (6:43:953) to (6:44:646)
Blue robot of blue team kicks the ball from(6:44:118) to (6:44:184)
Blue robot of blue team moves away from the ball from (6:44:184) to (6:44:976)
Cyan robot of blue team moves away from the ball from (6:44:349) to (6:44:613)
Cyan robot of blue team goes ball's Y direction from (6:44:613) to (6:44:811)
Blue team goal!!!! (6:45:42)
...

results for the first thirty five games of playing against a robot soccer team controlled by human operators using joysticks.

Table 10.5. Number of New Actions Executed in Each Game by a Team Playing Against a Human Controlled Team

Game	Total	Random	Other	%
1	3017	3017	0	0,00
3	4212	4168	44	1,04
5	2997	2973	24	0,80
7	3234	3162	72	2,23
9	3058	2886	172	5,62
11	3035	2812	223	7,35
13	3619	3464	155	4,28
15	3157	3071	86	2,72
17	3835	3701	134	3,49
19	4624	4497	127	2,75
21	4587	4422	165	3,60
23	4369	4369	0	0,00
25	3601	3447	154	4,28
27	4517	4347	170	3,76
29	3437	3346	91	2,65
31	4779	4779	0	0,00
33	3439	3285	154	4,48
35	3537	3454	83	2,35

It was observed that the team using the proposed approach had a positive and continued evolution in the number of new actions executed (e.g. non-random actions) during the first ten games; after that, evolution was slower but with increasing trend. It was observed that the trained team had a low rate of using new actions. It occurs because all states have the action "moves randomly" and the possibility of learning randomly actions. This happens because when the inference machine does not recognize a group of movements as an action in the observed team, then it forces to recognize them as a "moves randomly" action.

Another fact influencing the number of executed random actions is the action selection method when using reinforcement learning. Finally, it is important to note that, due to the exploration process in reinforcement learning, each action (including de random action), needs to be used many times.

In the proposed approach, knowledge base of each role starts without any state and only with the action "moves randomly". It is important to note that the number of known states increases during the game and during the training, on the other hand, the number of known actions increases only during the training.

Goals can be the result of a direct action or an indirect action of one's own team or a mistake of the adversary team. Since the main goal of a soccer team is to make goals, in order to evaluate learning capacity of the proposed method, an analysis of the number of effectively achieved goals (e.g. result of a direct action) was made. By using the computer program developed to narrate soccer games and analyzing its output, it was possible to determine if a goal was produced by a direct action of one's own robot, or by an indirect action or a mistake of the adversary team. The obtained results are shown in figure 10.9. Results shown that this approach allows the continuous evolution of the robotic team. This evolution is reflected in the increasing number of effective goals achieved by the learning team.

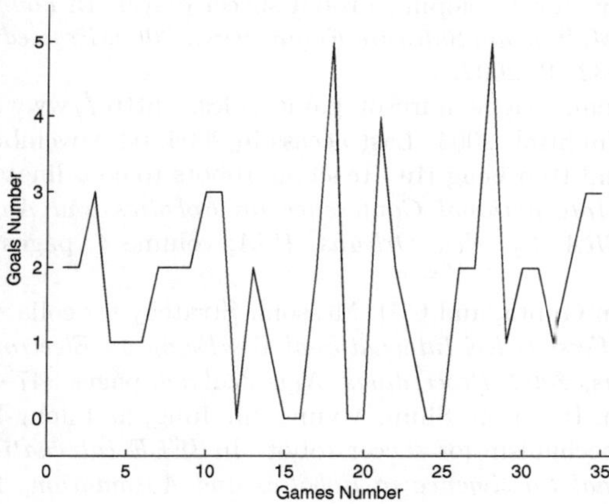

Fig. 10.9. Number of Goals Produced by a Direct Action

10.6 Conclusions and Future Works

This work shows that it is possible to recognize behaviors patterns in a micro robot soccer game using fuzzy logic and self organizing maps neural networks. This work is a base for an imitation learning by observing robot soccer games. Also was shown that an accurate robot soccer narrator can be developed using fuzzy logic.

This work was implemented using an architecture to control a robot soccer player with a dynamic strategy by mixing reinforcement learning and imitation learning. This approach was tested by playing with a static strategy and with robots controlled by humans and it reveals as a good way to implement robots that evolute with time and experience.

References

1. Dennis Barrios-Aranibar. Estratégias baseadas em aprendizado para coordenação de uma frota de robôs em tarefas cooperativas. Master's thesis, Electric Engineering - Federal University of Rio Grande do Norte, 10 2005. Available in http://www.ppgee.ufrn.br/teses.php.
2. Darrin C. Bentivegna, Christopher G. Atkeson, and Gordon Cheng. Learning tasks from observation and practice. *Robotics and Autonomous Systems*, 47(2-3):163–169, 6 2004.
3. James Bruce, Michael Bowling, Brett Browning, and Manuela Veloso. Multi-robot team response to a multi-robot opponent team. In *IEEE International Conference on Robotics and Automation, 2003. Proceedings. ICRA '03, Taipei, Taiwan*, volume 2, pages 2281–2286, 9 2003.
4. Kuan-Yu Chen and Alan Liu. A design method for incorporating multidisciplinary requirements for developing a robot soccer player. In *Fourth International Symposium on Multimedia Software Engineering, 2002. Proceedings, California, USA*, pages 25–32, 12 2002.
5. FIRA. Fira small league mirosot game rules. http://www.fira.net/soccer/mirosot/rules_slm.html, 2004. Last access in 2004, 04 November.
6. Dongbing Gu and Huosheng Hu. Teaching robots to coordinate its behaviours. In *2004 IEEE International Conference on Robotics and Automation, 2004. Proceedings. ICRA '04, New Orleans, USA*, volume 4, pages 3721 – 3726, 4 2004.
7. Sng H.L., G. Sen Gupta, and C.H. Messom. Strategy for collaboration in robot soccer. In *The First IEEE International Workshop on Electronic Design, Test and Applications, 2002. Proceedings, New Zealand*, pages 347–351, 1 2002.
8. Heung-Soo Kim, Hyun-Sik Shim, Myung-Jin Jung, and Jong-Hwan Kim. Action selection mechanism for soccer robot. In *IEEE International Symposium on Computational Intelligence in Robotics and Automation, 1997. CIRA'97., Proceedings, California, USA*, pages 390–395, 7 1997.
9. Y. Kuniyoshi and H. Inoue. Qualitative recognition of ongoing human action sequences. In *IJCAI93 Proceedings, Chambery, France*, pages 1600–1609, 1993.
10. Yuval Marom, George M. Maistros, and Gillian Hayes. Experiments with a social learning model. *Adaptive Behavior*, 9:209–240, 7 2001.
11. Hiroyuki Miyamoto and Mitsuo Kawato. A tennis serve and upswing learning robot based on bidirectional theory. *Neural Networks*, 11:1331–1344, 1998.
12. Itsuki Noda. Hierarchical hidden markov modeling for team-play in multiple agents. In *IEEE International Conference on Systems, Man and Cybernetics, 2003., Proceedings, Washington, D.C., USA*, volume 1, pages 38 – 45, 10 2003.
13. Itsuki Noda. Hidden markov modeling of team-play synchronization. In Daniel Polani, Brett Browning, and Andrea Bonarini et al., editors, *RoboCup 2003: Robot Soccer World Cup VII*, volume 3020 / 2004, pages 102 – 113. Springer-Verlag GmbH, 8 2004.
14. Rafal P. Salustowicz, Marco A. Wiering, and Jrgen Schmidhuber. Learning team strategies: Soccer case studies. *Machine Learning*, 33(2):263–282, 11 1998.
15. Stefan Schaal. Is imitation learning the route to humanoid robots? *Trends in Cognitive Sciences*, 3(6):233–242, 6 1999.

16. Jochen Triesch, Jan Wieghardt, Eric Mael, and Christoph von der Malsburg. Towards imitation learning of grasping movements by an autonomous robot. In A. Braffort, R. Gherbi, S. Gibet, J. Richardson, and D. Teil (Eds.), editors, *Gesture-Based Communication in Human-Computer Interaction: International Gesture Workshop, GW'99, Gif-sur-Yvette, France, March 1999. Proceedings*, volume 1739/1999, chapter 73. Springer-Verlag GmbH, 6 2003.
17. Marco Wiering, Rafal Salustowicz, and Jrgem Schmidhuber. Reinforcement learning soccer teams with incomplete world models. *Autonomous Robots*, 7(1):77–88, 7 1999.
18. Chia-Ju Wu and Tsong-Li Lee. A fuzzy mechanism for action selection of soccer robots. *Journal of Intelligent and Robotic Systems*, 39(1):57–70, 1 2004.
19. Marcelo M. Yamamoto, Antônio P. Araújo, Pablo J. Alsina, and Adelardo A.D. Medeiros. A equipe poti de futebol de robôs. In *I JRI - Jornada de Robótica Inteligente, Salvador - BA, Brazil*, 8 2004.
20. Marcelo M. Yamamoto, Diogo P.F. Pedrosa, Adelardo A.D. Medeiros, and Pablo J. Alsina. Um simulador dinâmico para mini-robôs móveis com modelagem de colisões. In *SBAI 2003 - Simpósio Brasileiro de Automação Inteligente, Bauru SP, Brazil*, pages 852–857, 9 2003.
21. Marcelo Minicuci Yamamoto. Planejamento cooperativo de tarefas em um ambiente de futebol de robôs. Master's thesis, Electric Engineering - Federal University of Rio Grande do Norte, 2005. Available in http://www.ppgee.ufrn.br/teses.php
22. Yu Zhang and Alan K. Mackworth. A constraint-based robotic soccer team. *Constraints*, 7(1):7–28, 1 2002.

Index

Author Index

MIX
Papier aus verantwortungsvollen Quellen
Paper from responsible sources
FSC® C105338

If you have any concerns about our products,
you can contact us on
ProductSafety@springernature.com

In case Publisher is established outside the EU,
the EU authorized representative is:
Springer Nature Customer Service Center GmbH
Europaplatz 3, 69115 Heidelberg, Germany

Printed by Libri Plureos GmbH
in Hamburg, Germany